LOST!

A Ranger's Journal of
Search and Rescue

A Graphicom Press First Edition

Includes Text, Line Drawings and Maps
Book Layout by Form + Function

ISBN Number 0-9641734-1-7
Library of Congress Catalog Number 97-073123

GRAPHICOM PRESS, INC.
P.O. Box 6
Yellow Springs, Ohio 45387

LOST!

A Ranger's Journal of
Search and Rescue

Dwight McCarter
Ronald G. Schmidt

Graphicom Press, 1998

Table of Contents

ACKNOWLEDGEMENTS

The authors and publisher wish to acknowledge the assistance of those who have helped bring this work to fruition. The illustrations are the work of Barbara S. Archer; layout and cover design by Eric Schmidt of Form and Function; editor Rita R. Colbert; assistant editor and administrative assistant Sharon Hatfield. In addition to his work as Marketing Director for Graphicom, Bill Scott served as field assistant, quartermaster and overall support staff to the authors. Annette Hartigan of the National Park Service assisted with research, as did some of the families of those lost. Cover photo supplied by the National Park Service which also permitted use of their trail map. Back cover photo courtesy of Clarence Maslowski. Our greatest debt is to our own families who have lovingly supported all of our efforts for many years.

DEDICATION

This book is dedicated to the families of those lost in the Great Smoky Mountains in sympathy with their pain and grief, and with the hope that the telling of their stories will bring some measure of peace.

FOREWORD

The Appalachians are venerable patriarchs among the mountains of North America. They have existed for hundreds of millions of years since the dawn of geologic time when they were created by primeval forces greater than our ability to measure or even comprehend. Even though the mountains are worn to only a shadow of the soaring mass they were then, their present soaring peaks reach northeastward nearly 2000 miles in an unbroken chain from Alabama to Maine and ultimately to New Brunswick and the Atlantic Ocean. At the ancient core of the Appalachians is its geologic and aesthetic heart, the Great Smoky Mountains.

Here and there the mountain barrier is breached by major rivers which have eroded their way through the chain, reducing the high peaks to ridges and valleys which are less impressive only by virtue of contrast with the highest reaches of the mountains. It is here, between where the French Broad River and the Tennessee River have cut through the mountains that our story begins. This is the place where the peaks are highest and the coves the deepest. Rounded hills and rugged ridges have been formed by the weathering of their ancient rock foundations, sculpted by the erosive power of streams, and now blanketed by trees which soften their knife edges. The beauty of the Smokies has been preserved as a national park for us to enjoy.

For many years our pioneer forebears saw the ridges and peaks as inhospitable barriers to the plains of the midwest. There the settler could clear level productive land and earn subsistence and, perhaps, prosperity by tilling the soil or trading with neighbors who were easily accessible. The products of his family's labor could be brought to market economically by flat boat or skiff on the navigable rivers. The weather was mild and friendly to his efforts. In contrast, the mountains with their unnavigable streams, steep slopes, impenetrable forests, and hostile weather all served to prevent migration and communication, and to ensure isolation of the settlers.

Eventually, the valleys were cleared and roads built to serve growing agricultural communities in the lowland areas, but the slopes and peaks resisted access until the industrial revolution brought loggers and railroads at the turn of the twentieth century. Subsequently, under the stewardship of the National Park Service, the natural splendor of the mountains has been restored and preserved for our enjoyment.

Today, many see the Smokies as an island of peace and tranquillity in a roiled urban sea, a refuge for those seeking communion with nature and the virtues of a simpler time. But the natural beauty and thin veneer of security and safety provided by a benevolent park service masks a more fearful side of the mountains. The very same characteristics which were hostile to the early settler are also hostile and often life-threatening to the unwary visitor today.

We humans lack the natural abilities to protect ourselves from the weather in the wild. By studying outdoor crafts and buying the latest high-tech survival gear, we seduce ourselves into thinking we are fit to battle any conditions in the wilderness, an illusion which sometimes costs us our lives.

This is a book about survival. It contains the stories of 11 people, all of whom challenged the mountains, fully intending to return to their homes safely. Some did and some did not. By telling their stories, we hope to increase the chances of survival of anyone else who comes close to the edge, or who faces the raw terror of being lost in the mountains. It is also about search and rescue, the sometimes rewarding and sometimes tragically disappointing effort to find and help the missing, no matter how brutal the weather or rough the terrain.

No one who has not searched for travelers lost and, perhaps, injured in the mountains can experience without a chill of apprehension, the rapidity with which night falls there, the sharp temperature drops, the updraft/downdraft winds which change from hour to hour, or the abrupt changes from sunshine to storm.

Spring and fall are particularly treacherous times in the mountains because the weather changes so quickly and frequently. Even in summer, without adequate protection, a quick thunderstorm can reduce body temperature sufficiently to induce hypothermia and even death. The symptoms are subtle at first and they are accompanied by a strong psychological anomie which dulls the sense of self-protection. In later stages, the skin burns as it freezes, and the sensation induces the victim to strip off gloves and boots and sometimes even clothing, accelerating the process. This bizarre behavior can be so extreme that some victims have been found nearly nude.

When someone is reported lost in the mountains, a complex process of search and rescue is begun. The task of searchers is a demanding one. They experience a roller coaster of intense physical activity distorted by intense emotional stresses. The hopes and fears of friends and relatives of the lost accompany the searchers, who bear the responsibility not only of finding lost loved ones, but of doing it before it is too late. Thus the urgency of the search hangs over everyone like a pall, and anxiety streams from eyes as tears. When the task is done, there is no other joy like that experienced when a child has been found in time after days of search. And no despair equal to that of finding one too late.

During Dwight McCarter's over 25 years of service to the park, much of it as the "back country" ranger, he participated in most searches. The stories in this book are all true. They come from the personal journal he kept throughout his career, the text of which forms the heart of each story. They are offered here with the hope that the lessons learned from them will help hikers understand how easy it is to lose their way, and if they do, to help them survive the ordeal.

LITTLE BOY LOST • ABE CAROL RAMSEY

A mile or so up the valley on Dunn Creek, just at the boundary where Sevier County meets Cocke, there is a clearing in the ring of the encircling mountain forest. In it, a crude mill stands close by the creek and, just beyond, a barn and a cabin. The slender cabin yard extends to an open pasture sloping down to the creek. At one end of the pasture is an orchard with gnarled trees, in this season bearing small, misshapen fruit and leaves just tinged with gold. A black horse and several brick-red heifers graze indolently in the summer heat. Flies buzz, circle and swoop around the animals whose skins quiver and tails lash at the intrusion. On higher ground beyond the house crouches a log barn with later additions made of clapboards leaning against the solid structure. The mill is below the house at a place where the stream runs rapidly downslope over large boulders. It is a small box of a building with peaked roof and sides made of plain rough-sawn boards nailed vertically and, like the house and barn, now weathered to a silver-gray color by sun and rain. By their look, none of the structures has ever seen brush or paint. It is early September of 1919 and this is the homestead farm of the John Ramsey family.

Behind a rickety fence of bent and twisted pickets and rails, and a narrow door yard of hard packed earth, stands the house, a smaller building also of raw, weathered boards with a pitched roof and deep eaves. On the left side, there is a well-built chimney of rounded river stones and mortar. In front, a shady porch faces toward the mill and the creek. The porch is clearly the living room of the house, at least in this season. It is littered with a spinning wheel which is about 5 feet in diameter and lots of other household clutter. There are strings of beans drying, two or three homemade chairs or stools and a rocker, all worn smooth and shiny by homespun clothing and work-worn hands and bodies. Here sits the man of the house, at this moment idle, and close by, his wife and a daughter who together are engaged in domestic work, both on their knees. Other children are playing about in the sunny yard at the foot of the steps.

Inside the house there are two rooms in the main part of the cabin and a kitchen in an ell behind. The furnishings are meager, simple and spare. Three big, high beds in one room provide ample space for the entire family of six and each is covered by a different patchwork quilt of striking pattern and color. In the next room there are more beds where guests might sleep provided they could successfully evict the hens which have laid their eggs on them. Here and there are more chairs and a rough stand or two. Left on the floor by the baby, a gourd with rattling seeds inside rolls about making its characteristic dry rustle as steps on the flexible floorboards disturb its repose. In the corner, a jumble of clothing. In the second room, the floor and every other horizontal surface is covered with newspapers upon which are drying beans. Pegs on the wall carry still more clothes and on the crude chimney mantelpiece stands a paralyzed clock and a couple of obscure ornaments. From the center of the ceiling hangs a lifeless electric light bulb on a string, a strange pretension in what otherwise would be a thoroughly honest scene. The kitchen is as simple as the rest with a dry sink, water buckets, a rough-hewn table, pie safe and cabinet to shelter food from the flies.

John Ramsey is a small man, slight of build and somehow managing to look both wiry and frail at the same time. He gives his age as forty-one years, but some think he looks younger, perhaps because they remember his once youthful bearing. He is blue-eyed and fair-haired, with a slack chin and an incongruously high voice. His wife, Mary Jane, although care-worn, fading, and now nearly toothless, still carries the ghost of a comelier youth, her brown hair curled over a high forehead. There were seven children. The eldest, Ola, still at home, is 14 or 15. Slender and naturally graceful, she is constantly busy with household chores, attending to these with quick, sure movements which convey a sense of habitual purpose. There is neither direction nor coercion from either parent here as she goes about her business of cooking, washing, milking or tending the other children. It is clear that here family living is shared in every respect as the children mature. Two older daughters were married and are now gone and Ola has simply taken their place as will the younger children when it is their turn and she herself is gone. She moves about the homestead whistling and singing quietly, apparently content with her lot.

Ruthie, just a bit younger, can be helpful too, but just now suffers from a bad tooth which makes her cranky. Next in line is Joe, a four or five year old tow-headed, nearly toothless urchin with a wry grin in ragged short pants and floppy shirt running about without a care in the world. The baby, Ida, not yet two years old, is whimpering and unhappy about something.

There is a gap in this family between Joe and Ida. And none of the others here can fill it no matter how hard they might try.

On the way along the old lumber road or any of the other trails or byways which converge on this place for miles in all directions, a stranger might stop to read a poster not long since nailed to a tree. It says: $100 Reward for Information leading to the Recovery of my Lost Child and is signed by one John Ramsey. And if anyone should ask of this gaunt mountain man "Would it be you who has lost a child?" "I be the one," he would answer quietly.

They are simple and kindly people, these Ramseys, as the mountain folk are widely known to be, and hospitable as well. If you come to their home you will be received with kindness and simple ceremony. A cool drink of branch water or cider will be offered, a bit of corn bread and grape jam or just a place to sit in the shade. Yet there is a sadness on their faces and in their voices which darkens their welcome whether to stranger or kin.

The shadow of their loss permeates the cabin. It shades and chills every activity. A conversation begun quite apart will inevitably veer toward it as if to a magnet. To any visitor who does not already know the essence, the story will soon be told. "One day last March, the eleventh day, about three in the afternoon," and it is true that in important things people are notably precise, "little Abe wandered away from the cabin."

"He was a fearless little feller," says his father, revealing by his words his puzzlement of the event and by his use of the past tense both his fear and his conviction of the finality of their loss.

And from then to now there is no sure sign that anyone has seen him. Neighbors and friends, even strangers from distant homesteads turned out to help. Every inch of mountainside, every wood, every rock, every crevice and every brook has been searched, far beyond the distance any three-year-old could stray alone. And nothing has been found but, some miles away, only some footprints of a booted man and with them, not continuously, but here and there along the way, those of a barefoot child. None can say whether or not the child was Abe or even if he was with the man or only coincidentally wandering along the same route. But there is more to the story .

A short while ago, and for the first time in the months since the disappearance, the family heard from beyond the mountains of a half-breed, half Indian, half Negro, who was seen near Soco Gap with a very young white child in his company. John Ramsey was immediately in chase. "One hundred and sixty-four miles," he tells listeners, "in three days," perhaps a bit of an exaggeration, but who could blame him for it?.

However, he returned empty handed; of the man, there was no trace. But Ramsey is convinced by what he heard, that the child was Abe. He talked with those who had seen the two and they described the child and his clothing and, to John's wishful thinking at least, the description jibed exactly with his knowledge. A woman, "She favored ma wife some," says John, told him how the boy ran to her, threw himself in her lap crying "Mama," and begged to stay with her. But the man, John was told, "set great store by the chile," saying "the little feller had twined hisself inter his heart an' he druther lose his han' as give 'em up."

And there was even more. The man had told others that he was on his way to West Asheville, where his wife's people lived, and that he would go from there to Pittsburgh, to the home of "Ider Shook", his sister. Later, when others who had heard the story turned the tale over to the police, there was a discouraging report. No alert had come from the sheriff of Sevier County, Tennessee; neither the name Ida Shook nor any close to it had been found in the local directory nor was it known to the department in Pittsburgh. No further clue has been discovered.

It is less than a mile from John Ramsey's cabin down Dunn Creek to his brother's store. Lloyd Ramsey's establishment is the community center; the rendezvous of these mountain people, a lounging place of the idle or expectant and a resting place for the traveler. It serves a basic need for communication, for fellowship, and for mutual support. It also sells a wide variety of articles and serves as the post office.

Loungers and onlookers pass the time of day, smoking, snuff-dipping and gossiping. Mountaineers ride up on horses or mules, attend to their business and ride away again. Lloyd Ramsey's young but knowing boy tends store in his absence. There are many examples of country wit and wisdom to be heard and enjoyed. A typical example: "When he dies," says one of the gathering of another, not present, "they kin truly say, 'here lies the truth,' fer hit never came outen him."

Years after Abe vanished another explanation was rumored about his disappearance. It was said that he wandered away from the house and came to an area where there was a whiskey still operation. No one was there at the time, but there were barrels of liquid buried in the ground and covered with shrubbery and leaves to conceal them. When the operators of the still returned, they made a frightful discovery. The Ramsey child had fallen into one of the hidden barrels and drowned. They knew of the Ramseys and their missing child, but fearing both the law and the retribution of John Ramsey, they said nothing, choosing to conceal the truth. They placed the tiny body in an old suitcase and buried it nearby. They removed the still and erased all evidence that it had been there. They may also have been responsible for creating and circulating the rumor about the mysterious tracks of the booted man and barefoot child which had been found three miles away from Dunn Creek.

This apocryphal tale is lent some credence by the fact that Lloyd Ramsey's store was known in those days to be not only the source of most of the supplies needed to distill whiskey, but also the place where the thirsty traveler could slake his thirst with something more astringent than branch water or buttermilk. On the way to the store, an observant traveler might well see a wagon drawn up with a full bed of straw beneath which, almost completely hidden, would be glass jars and bottles of an almost clear, faintly yellowish liquid moonshine.

The necessity of carrying the supplies for the still to that location and bringing back the finished product for distribution and sale would clearly require some location relatively close to the store. Thus the credibility of this legend is enhanced by the fact that such facilities were known to exist in the area and that they were probably nearby. If the tale is indeed true, it would have been a tragic irony that the illicit traffic of John Ramsey's own brother might have been indirectly responsible for the disappearance of his child.

In any event, little Abe Ramsey has never again been seen nor has any sign of his ultimate fate been found. He, like a dozen others in the history of the Smokies, has simply vanished without a trace. However doubtful it might seem, the conviction of his parents that he was kidnapped is a not uncommon reaction to such an event. It is very difficult for those close to such a tragedy to go on without closure. The absence of sure knowledge of death makes one search for and hold tightly to a more hopeful explanation, no matter how unlikely.

Source: A Hiking Trip Through the Great Smoky Mountains by Bayard H. Christy

FATHER'S DAY • DENNIS MARTIN

It was shortly after dawn when they drove up the Laurel Creek Road into Cades Cove. The trip from Knoxville had been uneventful and as they got out of the car and stretched they could see the summer morning mist still hanging low in the meadows of the cove. Parking in the picnic area parking lot was easy at this time of the day even on a holiday weekend in June 1969. The Martin family unloaded the camping gear from their car, checked their packs, and started up the Anthony Creek Trail. Although it was Friday the 13th, no one seemed apprehensive this warm summer day.

The Martin clan had been coming to this part of the mountains for years. In fact, this trip was an annual family ritual which had its origins many generations earlier. The tradition went back to the turn of the century when the Martin brothers, John and Jim, ran sawmills in the Anthony Creek drainage and on the mountain above. It was then that members of the family began the annual cattle drive from their farm in the Little River drainage near Walland, along the river, up Dry Valley to Schoolhouse Gap, across the Laurel Creek Valley to Bote Mountain near Cold Spring Mountain and eventually to Spence Field. Here the cattle would graze all summer on the lush grass, fattening for the next winter until fall, when they were driven back to the farm. Mostly they fended for themselves except for a single visit in June when the family brought salt for them. That June visit became the ritual family retreat to Spence Field after the park was established and cattle were no longer permitted to graze anywhere in the park. It was a particularly wonderful time to visit the mountain meadow because the azalea and laurel were usually in full bloom. Typically, about ten or 12 of the Martin Clan observed the Father's Day retreat tradition with great enjoyment.

This year was special because the immediate family composed of Clyde Martin, his son Bill Clyde, and Bill's nine year old son Doug, were being joined for the first time by six year old younger

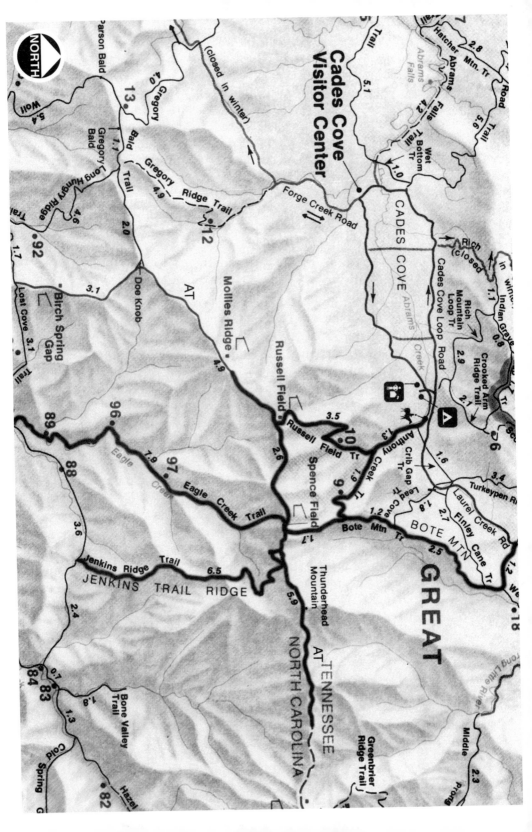

son, Dennis. Although this was his first overnight in the mountains, he was an enthusiastic hiker who had spent many days with his family on day hikes. He was a winsome child, strong and wiry, with a ready smile, curly brown hair and dark brown eyes, and a quiet but happy disposition. Dennis was eager to begin his first real camping trip in the mountains. Later the Martins would meet Clyde's brothers Doyle, Bob, and Huse, sister Irma and members of their families up at Spence. Right now they were joining distant cousin Carter Martin, who had come up from Alabama with his two sons. The families walked up the Anthony Creek Trail to the Russell Field Trail to Russell Field where they planned to spend the night. The youngsters, especially Dennis, were very energetic and often led the way on the long, sometimes steep climb to Russell. Like Spence, Russell was an open meadow which was starting to fill in with small trees and bushes since grazing had been stopped.

The group of seven spent the night at Russell without incident, rose early the next day, had breakfast, and hiked the short distance to Spence to meet the rest of the clan which had been there since Friday, Clyde's brothers Doyle, Bob, and Huse, and their sister Irma. Bob's wife Nita and Doyle's daughter Rita were the only other women who had elected to come this year. Bill Clyde's wife, Violet, had chosen to stay home with Michael and Sarah, Doug and Dennis' younger brother and sister. Other women in the family were similarly home bound with children who were too young for the trip. The walk to Spence was uneventful except for the humid heat and the pesky insects. Gnats and flies swarmed everywhere, especially in the open park-like meadows of the balds. Even the elevation didn't seem to reduce their numbers. Most of the wildlife seemed to be keeping out of the heat except for a scruffy yearling bear which would appear from time to time, snuffling and snorting. In the next day or two, members of the clan would see not only this animal again, but a young sow with two cubs, several times. Because they seemed poorly fed and bolder than usual, their sudden appearances made some members of the party uneasy.

When all the families joined at Spence, they prepared a mid-day meal, sharing the cooking and cleaning chores. By the time Clyde had finished washing the dishes at the spring with Dennis's help, it was about 2:00 P.M. Some of the adults were scattered near the new shelter at the southeast edge of Spence. Clyde, Bill Clyde and one or two others found a grassy knoll on the North Carolina side of the Appalachian Trail (AT) near the new shelter and stretched out to relax and enjoy the view of Rocky Top and Thunderhead Mountain. The Carter Martin boys who were about Doug's age joined Doug and Dennis to play in the field about 3:00 P.M. They ran around for awhile and then huddled, looked over toward the adults, and then split up. Doug and the two other boys headed south toward the North Carolina side of the field. Dennis, alone, headed west on the Appalachian Trail in the direction of Russell Field. To the lounging adults, it was apparent that the "gang" was planning to circle them and launch a surprise attack from behind. Sure enough, not long after, the three who had circled to the south jumped up, yelling and whooping as if they were part of some raiding party in the old west. But Dennis did not appear. At first everyone thought he was slower to get there because he had a longer way around to circle the knoll. When in a few minutes he still did not put in an appearance, the adults and the rest of the boys started to call to him. No answer. It was now after 4:30 P.M..

Slowly, feelings of concern and then apprehension began to steal over the group. Clyde and Bill Clyde walked quickly in the direction they had seen Dennis disappear only minutes before, but could not see him. Everyone continued to call, now as loud as they could, but there was no answer from Dennis. Tacitly, the men split up and each went in a different direction in search of Dennis whom they now assumed not to be answering because he had wandered out of earshot. Doyle went east and then south toward Haw Gap on the Jenkins Ridge Trail. Although he lost his way before he got back to the shelter, he was able to determine that there were no fresh tracks made by Dennis's shoes in the soft or sandy patches near the trail. Clyde made a quick sortie to the east end of the field and then along the AT toward Rocky top, also without any sign of Dennis in that direction. Bill Clyde hiked very quickly west, first to Little Bald. When on his return, there was still no word of Dennis, he went west again, all the way to Russell, calling and looking for signs of the boy. Hikers he encountered at Russell who had walked up from Fontana Dam on the AT reported no sign of Dennis along the western end of the AT.

Meanwhile, the level of concern among the rest of the family at Spence was reaching a high pitch. Everyone was searching for Dennis, calling out and walking around Spence Field looking for the boy. His grandfather, Clyde, hiked down Bote Mountain Trail to Anthony Creek Trail and then down to the ranger station at Cades Cove where he reported Dennis lost at 8:40 P.M.. Once again he climbed the long trail back up to Spence. While he was gone, Bill Clyde returned from Russell and hiked down Bote Mountain Trail toward the Laurel Creek Road. Along the way he encountered Terry Chilcote, the park naturalist, and his wife who had driven up the trail. When they learned of Dennis's disappearance, they drove Bill Clyde down to Laurel Creek Road where they left him by the side of the road. In a short time, Ranger Larry Nielson came by from the Cades Cove station in his jeep and drove Bill Clyde back up the Bote Mountain Trail to Spence.

Ranger Nielson's first response to Clyde's report had been to dispatch Ranger Dennis Huffman from the Cades Cove station to hike up Anthony Creek and Russell Field Trails to Russell and then east along the AT to Spence where they would meet Nielson who drove down Laurel Creek Road to the Bote Mountain Trail crossing and then up to Spence, having picked up Bill Clyde on the way.

By this time it was after 9:00 P.M.. The extremely hot and humid weather of the past few days broke quickly, starting with distant rumbling thunder to the west and a freshening breeze. Within a matter of minutes, the trees were swaying and scattering leaves; dark clouds poured in and sheets of cold rain drenched the field and the searchers still calling for Dennis. Realizing that their calls were lost in the noise of the storm, and that they stood little chance of even seeing the boy in the rapidly falling darkness, they gathered in the shelter to wait out the deluge. Although the cool rain was at first a relief from the sweltering heat, they all chilled rapidly and that realization gave rise to another concern about Dennis - hypothermia. A little boy with only a tee shirt and shorts could be expected to lose body heat and chill rapidly without waterproof clothing or shelter. As the night wore on the big storm continued to rage outside the shelter and those within were anguished in their own sleepless thoughts of a lost little boy.

THE JOURNAL

Saturday, June 14, 1969 10:30 P.M. - Day 1

I just finished reading the paper and was watching and listening to the storm for over an hour when the telephone rings. The call is from Chief Ranger Snedden who tells me about the Martin boy and asks me to report for duty tomorrow at 5:00 A.M. at the Bote Mountain Trail. I do not say so to him, but I know the rain storm is a major complication to the search, greatly diminishing the chances of finding Dennis. There a serious risk of exposure and hypothermia for the boy. The rain will wash out whatever tracking sign we might have been able to find, rendering the use of tracking dogs all but useless. I set my alarm for an early rise, but that turns out to be a joke. Sleep is first elusive and then fitful when it finally comes.

Sunday, June 15, 1969 5:00 A.M. - Day 2

I am up before the alarm goes off and at the rendezvous early, assigned as a member of a four-man search crew going to Spence. Getting there is the problem. Although there are nine jeeps and three trucks available today, every jeep we can find is already in use to shuttle searchers up to Spence. The last mile or so near the top is the worst problem because each vehicle headed up meets others coming down and there are few places to pass. Over 2.5 inches of rain fell last night in the storm. All of the streams are up and very turbulent with most over their banks. The trail-road up Bote Mountain is in bad shape with running water, mud and washouts everywhere. It takes a long time to get to Spence.

After arriving at the field we are sent west to Mount Squires and there we turn down Anthony Creek Drainage and search downstream. Later we tie into the Anthony Creek Trail and walk back up to Bote Mountain Trail where we catch a ride back to Spence. There is no sign of Dennis in the area we searched. At the top we do the same thing again in the same drainage but our starting point is dif-

ferent. We have the same results, nothing is found.

It is almost noon now and we return to Spence Field where a base tent has been set up. We get in a long line at the tent to pick up lunch; a sandwich an orange and a drink. Spence Field is teeming with searchers, over 150 in all. Jeeps are coming and going bringing searchers up the Bote Mountain Trail. Two helicopters arrive and ferry people to different locations. The search seems to be using the same type organization as you would have on a major forest fire.

1:00 P.M.

We are given a new assignment to join with approximately 30 other searchers. We start a grid search from Point Last Seen (PLS) and traverse in an easterly direction on the Tennessee side going toward Rocky Top and Thunderhead. It is slow going as we are 15 feet apart and have to realign the search line often. The area is steep, rugged, and choked with laurel. We hear other searchers on the North Carolina side doing the same thing. At Thunderhead Mountain we pull out to the AT and hike back to Spence. It's getting dark.

As we pass by Rocky Top, I notice some old carvings on an exposed rock about ten feet into North Carolina. The inscriptions read, "Hop Harris" and "Red Waldron" with dates they visited here in 1791, 1794, 1797, 1799. It then skips to 1812, 1813, 1814. I imagine the reason for the gap in visits was because the then Governor John Sevier was involved in the Cherokee Indian Wars from 1800 to 1812 and the area was not safe.

We are flown down to the cove via helicopter. All trails on both the Tennessee and North Carolina sides of Spence were walked and re-walked this day. There is no sign of Dennis. A 24 hour watch is installed at Spence with the family remaining there at night.

Temperatures are in the 50 to 60 range at night and in the 80s during daylight. There were, officially, 240 searchers looking for Dennis Martin today.

Monday, June 16, 1969 6:00 A.M. - Day 3

We are driven to Spence via jeep. This day we are members of a large grid search that proceeds from Point Last Seen west toward

Russell on the Tennessee side of the Appalachian Trail. We swing around past Little Bald and grid back toward Spence. There are numerous old camps at Little Bald. We reach Spence at 1:00 P.M. After lunch, our team continues gridding on the North Carolina side of Spence going toward direction of Rocky Top. We find many old backpacks, binoculars, cowbells, etc.; most are things that have been dragged off by bears from past camps on Spence. I hardly catch a glimpse of my old friends Larry Nielson and D.K. Huffman as they are busy dispatching teams and coordinating the intensive search. Likewise I have only occasional meetings as we pass trackers J.R. Buchanan, Art Whitehead, and Grady Whitehead as they are assigned to hot areas where some sign may have been found. They are mountain men of the first class. I know that if they can only find the boy's tracks, Dennis will be found.

The north shore of Fontana Lake is being searched and patrolled both on foot and by boat. There are news people everywhere and the search is beginning to look like some kind of side show with cameras and reporters shooting film and interviewing people.

1:00 P.M.

In no way do I possess the man-tracking skills that Buchanan and the Whiteheads have but I have been lucky enough to have accompanied them on different searches in the past. I am hopeful that I can apply some of the knowledge gained from them in this search as we continue our grid east toward Rocky Top. We cross the trail to the Old Spence Shelter. Further on we cross the Jenkins Ridge Trail many times as it is going the same direction as we are. The terrain opens up a little as we near the Devils Race Patch and Meadow Gap. I have the feeling the Meadow Gap area must have been occupied in the long ago past as there is fairly open forest, a spring, and other signs. Maybe this was a hunters' camp area or before that an Indian camp. We have no luck in our rapid grid search and returning to Spence, we encounter several dog teams returning without contact with Dennis. He seems to have dropped off the face of the earth without a sign. It is impossible to track when there are none. We ride jeeps back down to the cove as it gets dark.

As the scope of the search is increasing, so is the number of searchers. There are reported to be more than 300 here today with personnel from the park, local rescue squads, the Air National Guard, and just plain volunteers who have come in because they heard about it from the media.

Tuesday, June 17, 1969 6:00 A.M. - Day 4

We arrive at Spence once again. The overnight low here was 53 degrees but it already feels like it will be another very warm day today. The Martin family, including Dennis's mother, have maintained a continuous vigil at the field. She was not on the camping trip with Bill Clyde and the two older boys but she came on when she learned he was missing. Dennis's father says it seems like a dream that a child could disappear so quickly. He describes Dennis as shy, a quiet child who keeps to himself. I wonder to myself if that means he will not reveal himself to searchers. It sometimes happens that a missing person, especially a child, hides from searchers out of some primitive fear. If Dennis is one of those, we may never find him.

Small groups of search volunteers camped in the area last night, building large fires in hopes that Dennis might see them and be attracted. It rained again last night and the clouds hang low this morning on Spence and Russell inhibiting the helicopter shuttle. There are now 4 choppers here, but they are all grounded until almost noon. The Bote Mountain Trail/Road is a mess with constant traffic churning up the mud. Several truckloads of crushed rock have been brought in to stabilize the worst places, but it is still bad, slowing the transfer of personnel and supplies. Ninety seven people are assigned to grid searching the Spence Field area.

We grid the Anthony Creek Drainage again toward Cades Cove on the Tennessee side. We grid the the West Prong Drainage also on the Tennessee side. No luck in either area. I do find oranges and apples that have fallen from previous searchers coming through the area. Other grid searchers are gridding the Eagle Creek drainages and the Hazel Creek drainages on the North Carolina side of Spence.

Early Afternoon

The family said that Dennis had eaten some soup at about 2:00 P.M. on Saturday, the day he was lost. He is wearing a red tee shirt and green short pants with white socks and black oxford-type shoes with a bed. At approximately 2:00 P.M. Tobe Patterson and Jack Harmon find an indistinct footprint on the North Carolina side of Spence field. They are members of a group who were sweeping down a ridge at the headwaters of Eagle Creek Drainage. Both men followed a stream branch to the right of the creek for a half mile where they found a shoe print and a footprint which were about 7 inches long and spaced according to a child's stride. Jack says that the track is that of a child age 5 to 7 years old, but he thinks it is a tennis shoe not an oxford-type shoe that would definitely have a heel. Rangers make a casting of the print which the parents say is too large to be their boy's track. It is important to note there are many other children coming through the area both before and during the search. Another set of prints found Monday were also ruled out as they were too big.

One searcher accidentally shot himself through the right leg as he was unloading a pistol during the search earlier today. During this day of searching we have heard many gunshots far off and each time I wonder if the boy is found, if this is a way of communicating, or if there are more people with gunshot wounds to the leg or foot.

We are starting to hear about psychics and clairvoyants and their predictions for finding Dennis.

Hope flared this day with reports of a boy being found at Cades Cove. It turned out that the child had strayed on a fishing trip and had been found quickly. It was not Dennis. There is much confusion and the woods are full of people calling to one another. It is hard to know whether the sounds mean anything to our search. We complete our assigned area and return to Cades Cove as darkness is falling.

The weather has continued hot with thundershowers again. Three hundred and sixty five people took part in the search today including 149 from 20 different rescue squads, 40 Special Forces personnel from the military, 50 junior college students, 75 National Park Service personnel and 51 other assorted volunteers. There is now a helispot at Cades Cove as well as one at Russell Field with refueling and supplies transfer operations at Cades.

Wednesday, June 18, 1969 6:00 A.M. - Day 5

It rained again last night and the aircraft ceiling is 4000 feet,
below the elevation of Spence. We are now in the habit of riding up
by jeep in the morning and riding down via helicopter in the
evening since early fog hampers helicopter operations almost every
day. If it rains again today, the Bote Mountain trail/road may no
longer be passable and search operations may be seriously curtailed. This day we grid the Haw Gap area on the Jenkins Ridge
Trail in North Carolina. I especially like the Blockhouse Mountain
area and the De Armond Bald which are on either side of Haw
Gap. There is evidence of old camps at Haw Gap. At Blockhouse
Mountain there is a large meadow on top of a lone mountain.
There seems to be an ancient presence on this mountain, perhaps
from early Cherokee or even earlier Indians who left their intricate
carvings not far from here. From Blockhouse Mountain we drop
down Forrester Ridge into the Bone Valley of Hazel Creek and
proceed to Halls Cabin where an army Huey helicopter will pick us
up. We find no sign of Dennis on this day's search. Halls Cabin has
a small meadow below it and there is a spruce tree near the
meadow. The chopper lands O.K. and we load on, but on take-off
the rotor shears off the very top of this tree. There is no damage to
the chopper however and we land safely in Cades Cove as darkness
is closing in.

This evening 15 searchers are spread out at intervals up the
entire Eagle Creek Drainage. They will maintain fires all night in
hopes that Dennis will be attracted to them. It seems like the search
may be intensifying on the North Carolina side. Some think Dennis
may be in Eagle or Hazel Creek Drainage.
The total number of searchers today reached 615.

Thursday, June 19, 1969 6:00 A.M. - Day 6

This day I join a small team of three searchers and a dog. We
drop off the AT into Hazel Creek Drainage. I believe we are on the
Welch Ridge. I do know that we are dropped off by helicopter in a
meadow east of Spence on the AT so it has to be Silers Bald. Anyway
we walk down the entire ridge all day and are picked up on the
shore of Fontana Lake. I'm not very familiar with this section of the

park. We see plenty of animal signs but no sign of the boy.

This day there are lookouts stationed in High Rocks Tower that looks into Hazel Creek and at Shuckstack Tower that looks into Eagle Creek. They are to look for buzzards and plot their location. Also, all pit toilets at Spence, Russell, Mollies, and other locations with these type pits are to be examined. That means lowering a searcher into them. Fortunately I am not chosen for this detail! Today searchers are instructed to begin to collect any bear or boar excrement for analysis. A three mile radius has been covered with all ridge tops and drainages combed.

Psychic Jeane Dixon predicted searchers will find young Martin behind a waterfall on the North Carolina side of Spence Field. She said he went out level first, then went down an incline and turned left at a 40 degree angle and up a little, then went back down and would be below a point of incline. The left turn is in a clearing with no trees. It is more or less bare ground. Searchers following this advice as best they could failed to find any sign of Dennis.

Another psychic, Mrs. Schwaller of Linden, Michigan said the boy would be found five miles southeast from PLS on a stream by a waterfall and that white pine trees are in the area. Southeast from PLS is 135 degrees and following the compass it crosses directly over Halls Cabin on Hazel Creek and the five mile mark is at 3200 feet elevation on White Walnut Branch. Special forces checked this area also with no results.

Another prediction by Harold Sherman of Los Angeles, CA says the boy will be found two and one half miles to the left of PLS, where he fell off a steep place, hung up in bushes. The only problem is left from what? Some think that this location would be on the Mill Creek below Russell Field, but I don't know why. This area has already been searched to no avail. We hear that Harold Sherman has clarified his prediction. Assuming the boy to be standing facing west toward where his father is seated. He sees Dennis moving to the right then veering to the left. He would then be on the Tennessee side. He will be found two and one half miles from his father. He feels that a sex maniac could have trailed the boys. He sees some-thing red being carried or dragged. Are there any caves? Not a large one, and not far from a stream. He gets a feeling of sudden panic from Dennis. He did not cry out, he was struck down by something. The area is rugged. Searchers are sent to check out this prediction, but nothing is found.

The army specialists are doing another grid search between Forrester and Jenkins Ridges. We are picked up at Fontana Lake and dropped at Cades Cove as it gets dark. Six hundred ninety searchers combed the mountains for Dennis today. He will be seven years old tomorrow.

Friday, June 20, 1969 6:00 A.M. - Day 7

We are among 780 searchers today. We proceed to Russell Field and grid down the Mill Branch Area. At the lower end of the field we find an old mowing machine, the kind horses would pull.

Contingency plans were made today for procedures to be followed when Dennis is found. Under Plan A (found alive) the boy will be taken by helicopter to Knoxville headquarters of the U.S. Marine reserves and then by ambulance to University of Tennessee hospital. Under Plan B, the Blount County Coroner will be called in to take charge. We are given instructions for procedures when we find the boy:

1. Determine if dead or alive (dead only if rigor mortis has set in).

2. Notify Chief Ranger by most expeditious method and give location in detail, radio code 10-200 if dead or radio code 10-100 if alive.

3. Climb a tree and set a flag, build a smudge fire, use a smoke bomb (military only) or other signal for helicopter.

4. Stand by while special forces rappel a man in by chopper and secure boy in litter if alive or, if dead, guard area until released by Chief Ranger or coroner.

5. Get name and address of person or persons who found boy.

We hear that additional beltspots have been set up at Thunderhead, Derrick Knob, Gregory Bald, Eagle Creek, Hazel Creek and Fontana in preparation for massive search activities tomorrow and Sunday.

We finish gridding the entire Russell area finding the usual amount of stuff bears have drug off from campers in the past. We have no luck again today and walk down the Ledbetter Trail at nightfall feeling discouraged.

Happy birthday, Dennis, wherever you are.

Saturday, June 21, 1969 6:00 A.M. - Day 8

A roadblock has been established at the forks of the Little River or Townsend "Y" at 5:00 A.M. today to keep park visitors out of the search area and to control the large number of volunteers which are expected today. We travel by bus up Laurel Creek Road to Cades Cove where two Chinook helicopters and four other smaller choppers are airlifting searchers to the mountain beltspots. I have never seen anything like this before. With the large number of both military, park, and volunteer civilian searchers, and all of the vehicles and equipment to service them, it looks like either an invasion is being launched or an evacuation is underway. Many volunteers are agitated because they have to wait hours for transportation, first at the "Y" then at the beltspots and other transfer points. Many never get to the search areas.

Today I am assigned to assist rescue squads in a search of Thunderhead and Lynn Camp Prongs and the Derrick Knob area. While gridding Thunderhead Prong we search around the Thunderhead Rock and I photograph the old camp rock since we have found no clues. We also visit another camp rock on Lynn Camp Trail and come upon one of the largest rattlesnakes I have seen in awhile. It has just swallowed a squirrel which it coughs up when we disturb it.

As of today, a one mile radius around Spence is gridded and all ridge tops and drainages have been checked to 250 feet on either side. Rescue squads drive jeeps to Derrick Knob and we also search a large section of Middle Prong at a place called New World. This area is at the Upper Thunderhead Prong and is the site of an old logging camp and railroad from the 1920's. There are pits, barrels, old foundations and the old railroad grade still evident. New World is slightly uphill from the lumber era foundations. It gets its name from a settlement dating back to the Civil War when several families moved here to escape the war. They lived in peace through the conflict because they erased any sign of their coming and never made any contact with the outside world during the war.

One searcher suffered a broken right arm when he fell crossing a steel beam across the river. There was a terrible tragedy here in 1918. Two men, Sands and Davenport, were coming through the area from Spence Field. A freezing storm with heavy rain was in progress. Sands fell in Thunderhead Creek while fording it. He began freezing. They went to a local camp rock shelter but they could not get a fire started as matches were wet. Davenport quickly hiked 2 miles down river to the big hollow where a man named Moore lived. Moore brought his mule and accompanied Davenport back to rescue Sands, but found he was already frozen to death.

We found no sign of Dennis today. Psychic Jeane Dixon said Saturday the child was still breathing Friday night. An 18 mile radius has been covered in general terms. A six mile radius has been covered in specific terms, and a one mile radius from Spence has been saturated with detailed grid searches. We are told that there are 1400 searchers from 35 different organizations looking for Dennis today. Once again we make our way down from the ridges without even a sign of Dennis.

Sunday, June 22, 1969 6:00 A.M. - Day 9

The number of searchers today is less than yesterday, but there are still more people than we can use efficiently. Much time is wasted with personnel standing in line for everything from transportation and food to toilet facilities. Everyone is very frustrated and some who have been searching for over a week now are discouraged.

We are assigned to re-grid Anthony, Ledbetter Ridge and Mill Creek at Russell. I eat some green apples at Russell. This area was surely occupied in the pre-park days. I wonder if Dennis has had green apples to eat or anything. It is too early for berries and such to be ripe so I wonder what he might be eating. These grassy fields appear to be very old and I recall reading of an early explorer in 1740 who followed buffalo to this place and saw them grazing in a field two miles west of here. He said their well worn trails were 3 feet wide and followed the easiest route. Searchers in the field today are checking and double checking areas. We have again spent an entire day without finding any trace or even a clue as to what might have happened to the curly haired boy.

We learn that the search directors now believe that all of the likely locations in the park have been searched without success. A decision is made not to continue to expand the search area, but to begin again by combing carefully the areas close to Spence beginning tomorrow. Over 1000 searchers were involved in the work today.

Monday, June 23, 1969 - Day 10

There is heavy rain with flash flooding last night and today. Search operations are curtailed today since the helicopters are grounded and the Bote Mountain trail/road is nearly impassable. Neither the mountains nor the search need more rain. One dog team went out without luck. We are "dog tired" from only getting a few hours of sleep for many days and hiking strenuously up and down hill all day long with little little rest. There is, of course, a sense of great urgency to try to find Dennis. A little boy could survive on green apples or roots or even no food at all for a week or so as long as he has water. There seems to be no lack of that commodity. If he has been able to find shelter from the many cold rain storms and could stay relatively dry, he could be alive and maybe not even in dire straits. However, time and search possibilities are running out and we all know that after ten days, the likelihood of a tragic end to this search is growing more likely.

Many different theories about the disappearance of Dennis which could have impact on the search have been explored and discussed with each getting some attention. All seem to be some version of three basic ideas:

For over a year now the food supply for bear and other wildlife has been in short supply. Two years ago the acorn crop failed and many animals, like bears, which use this material, called mast, as an important food source have suffered. It has resulted in unusual mass migrations of animals both within and outside of the park. The bear have been particularly hard hit and have become very aggressive in their encounters with hikers and campers who are carrying food.

It was so bad that a strange experience I had then stuck in my mind. Just a week before Dennis' disappeared, I was assigned to work with the Cades Cove Trail crew. We arrived by jeep at the Spence Field and were planning on going to Russell Field and down Ledbetter Trail to the cove, clearing and repairing trail as we went. At Spence it is our custom to see if any bogs are in the trap located at the south side of the field near the wood line. We were surprised to find a bear in the trap. The bear had eaten all the corn that is used for bait, which is very unusual since bear do not usually eat corn. This bear was in extremely poor condition and in a very bad mood. One of my co-workers climbed atop the trap and raised the door whereupon the bear charged out, spun around a few times and ran off to the east toward Thunderhead. I thought the bear must be very hungry to eat corn. We proceeded on about our work that day without further incident.

A few days later, the local paper had a story indicating that last fall the mast was again a failure. Many wildlife could not find food in the form of nuts on the ground. This is the food that the bear and others depend on in the fall to add layers of fat for the coming winter; so in the spring of 1969 these bears were very poor. The paper indicated the reason for this mast failure was a very dry summer of 1968. I do recall having many encounters with bears in the backcountry this spring. These bears would bluff charge me in an attempt to get me to drop my food pack. My response was to charge them back and deliver a blow to their nose with my stick. This worked 99 percent of the time with the bear taking off. The other one percent the bear did not bluff and I would climb a tree until she left. I guess if my stomach had been empty for quite awhile I might take the risk also.

Even without knowing of these incidents, the search leaders explored the question of whether or not Dennis might have encountered a bear or boar and been killed by that animal. The possibility has been mostly discounted by the general knowledge that although such encounters are not really rare, almost never do they result in death or even injury. Black bear are usually not aggressive enough to kill large game or humans. They prefer to subsist on vegetation and found carrion. Even if a child could be killed in that way, the idea of there not being any remains to be found seems unlikely to many. They point out that bears usually do not eat freshly killed meat, but wait for it to ripen so a body would be likely to be found before it is consumed. On the other hand, once a bear takes possession of a dead animal, the remains are hidden and often covered with leaves or soil until they are ready to be consumed. So, if Dennis has met such a fate, his body, with his easily recognized clothing, might actually be hidden from the view of even the most sharp-eyed searcher or experienced tracker.

The family of the boy seems to favor the idea of Dennis having been kidnapped. This notion, suggested or reinforced by some of the psychic visions seems to be growing in favor among some of the searchers as well.

The third idea which seems to be the most popular and has been the central assumption of the search has been that Dennis became disoriented in his maneuver to surprise his family and somehow plunged deeper into the forest in some direction or other. If so, he has either been wandering around out of sight and sound of the searchers or has become injured and unable to respond. Since it is not uncommon for lost individuals, especially children, to hide from their searchers, it is even possible that Dennis has been eluding the search net.

So far there has been no direct evidence of any kind which supports one or another of these theories. Dennis has simply vanished so far without a trace. The impact of such a strange and mysterious event, if it turns out that way, would be stunning not only to his family, but the whole world.

There were 427 people looking for Dennis this day.

Tuesday, June 24, 1969 - Day 11

We return to Spence Field and again grid west toward Russell. Our search now has encompassed a 13 1/2 square mile area centered around the PLS. Today we widen the radius of search and re-entered previous checked areas. Two dog teams accompany us. Rainfall of 2.72 inches has fallen overnight and creeks are high. Bill Martin is using a megaphone to shout to the boy. One green beret lost his wallet with $600 in it on the Tennessee side of

Thunderhead. What is he doing with that much money in the woods?

Hope flared anew today as word came over the radio that a boy with red T-shirt and green short pants was spotted at Cades Cove campground. It turns out to be one Michael Devlin of McCune, Kansas who is staying in the campground with his family. Search leaders ask him to change his clothing for other colors to prevent further confusion, which he does.

We hear that the FBI has been called in to investigate the possibility of kidnapping. They have been questioning some people and following leads provided by the psychics. There is no information about any results.

Four hundred eighty two searchers continue looking for Dennis today.

Wednesday, June 25, 1969 - Day 12

We return to Spence via helicopter today. I am part of a group of 100 National Park Service (NPS) and volunteer searchers. We grid further east on the AT toward Derrick Knob. It is really a beautiful area on the east side of Thunderhead with birch and beech and open forest. However a short distance off the trail to the west you look down a precipitous bluff. At Beechnut Gap I go down to the spring on the Tennessee side and find old barrels, water pipes and old mounds of earth. This area has been searched earlier as I notice ample flagging present. Many of my fellow searchers still think Dennis is within a mile or two of Spence. I don't know what to think. I have just turned 24 and have yet to gain the woods savvy that most of the old timers have. We do not find the boy today.
The total number of people searching for Dennis today is 463.

Thursday, June 26, 1969 - Day 13

Most military personnel are pulled from search. We are 120 searchers today, mostly volunteers. We re-grid some areas east of Spence that had gaps in them. Nothing is found.
A press release is distributed which says that the search will be greatly reduced as of tomorrow. If the boy is not found by June 29, the search will be continued on only a limited basis for 60 days.
Dennis's father and family left the search Wednesday, June 25, and went home.

Friday, June 27, 1969 - Day 14

We are involved in trail cleanup in the Spence area. All military except one Huey pulled out today. There seems to be a deep foreboding feeling as the search winds down. Dennis's great aunt says they will be on Spence every day for the rest of the summer. She recalls that the family was within 30 feet of the child and saw him heading for the Anthony Creek Trail toward the Tennessee side. This is the last day I will be involved in the search. A great sense of failure and disappointment overwhelms me but I believe we have done our best with the knowledge, skills, and abilities we possess.

Sunday, June 29, 1969 - Day 16

Spence is shut down at 6:00 P.M. Everything is removed. Three of the park's best mantrackers remain on the search. They are brothers Art and Grady Whitehead and J.R. Buchanaan. The last helicopter is gone.

Between Day 20 and Day 66, as the search wound down, a strong smell of decay at West Prong reveals only a dead crow. A vulture circling the southwest end of Cades Cove was spotted; searchers found a dead bobcat. A hiker reported that on June 14, 1969, he and his sons had heard a child scream and saw a man hiding in the bushes. Officials say the distance was too great to have been connected with Dennis's disappearance. The Martins offered a $5000 reward for the return of Dennis and distributed posters offering a reward for information about Dennis in Gatlinburg and other towns in the area.

Dennis is still missing. In almost 30 years no definite clue as to how he disappeared or what fate might have befallen him has been found. Every possible lead has been pursued and none could be linked to him.

There remain three possible scenarios to explain his disappearance. The first and simplest is that when he tried to circle around the adults he got into the woods, became disoriented, and wandered further away from the others. He might have fallen, been struck by lightning in the rain storm which came shortly after he was lost, or otherwise injured himself, or been killed. It could even be likely that hypothermia from the night temperature and heavy rain, even in the warm days of early summer, was enough to take his life. Although one might think that the massive search efforts would render this possibility unlikely, anyone who has walked off the trail in these mountains knows that the ground cover is so dense in many places that the body of a small boy could easily be missed. If he took cover in a laurel or rhododendron thicket, he might be impossible to find. The only real evidence for this scenario however is the odor of decaying flesh we smelled on July 3rd and a report which only reached me years later.

In July of 1985, I was contacted by a long time ginsenger whom I know well. He related a story of one of his trips in search of ginseng in the park several years earlier. He said that he and one other person were looking for 'seng up the Big Hollow in the Tremont Area. They were proceeding up the right side of the stream and were 200 feet away from the creek. As they neared a little waterfall he noticed some bones lying near where a tree had uprooted and left a level place. He said the bones were that of a child which included the skull. He said it looked like animals had scattered the bones. He thought that this might be Dennis Martin. Not long after this report I contacted my very good friends at the Swain County rescue squad in Bryson City, North Carolina. They brought 30 men the following weekend and we did a thorough search of the hollow but found nothing.

The second possibility is that he might have been attacked and eaten by bear or boar. There are no known instances of death caused by either animal inside the Great Smoky Mountains National Park. There have been injuries, however, and a small child might be an attractive quarry for a large enough bear, especially if it were starving as they were that year. For the most part, bears do not eat fresh meat but prefer to consume flesh after if has partially decayed. Typically, a bear will partially bury or hide a found carcass of an animal until it has decomposed and then return to consume it. Although this could have happened to Dennis, especially if he died from other causes, there is no evidence for it other than the known presence of hungry bears in the area.

The third scenario is a kidnapping. Perhaps someone seized Dennis, carried him away, and has kept him captive and away from others, at least long enough for him to forget his early child-hood and to be acclimated to a new life. The only evidence for this possibility is the information provided by Harold Keys who heard a scream and saw a man hiding in the bushes in Sea Branch drainage near Cades Cove. Although the Martin family favors this explanation, the evidence is spare and the likelihood of maintaining isolation and achieving memory repression in a six year old child for such a traumatic event seems less likely. It is also a possibility which the FBI investigated at the time and could not find sufficient reason to pursue further.

THE PERILS OF PASSAGE • GEOFFREY BURNS HAGUE

On February 7, 1970, Scoutmaster Eugene Smith, leader of Boy Scouts of America Troop 95 of Morristown, Tennessee, his assistant, Marvin Horner, and Reverend Pitser Lyons started out on an overnight hike along the Appalachian Trail (AT) east of Newfound Gap with three teen-aged explorer scouts . They were Lee Smith 15, son of the scoutmaster; Steve Wolfe 16; and Geoff Hague 16, who had just earned his explorer badge in scouting. The group of three adults and three boys arrived at Newfound Gap in late morning, parked their cars, and hiked the three miles east on the AT to Icewater Springs. The hike in was routine. The weather was overcast and the temperature in the twenties. At this elevation, above 5000 feet, and in this midwinter season, there was at least five inches of snow on the ground, and more where it had drifted. The forecast was for similar temperatures and more snow the next day.

The group pitched their camp at the National Park Service (NPS) shelter at Icewater Springs. Later in the afternoon, Marvin Horner and Reverend Lyons returned to the Newfound Gap parking lot and headed for home. The boys and Scoutmaster Smith set about making their digs ready for dinner and the night. After a little while, the boys started a snowball fight which became a little too rowdy; Scoutmaster Smith brought the roughhouse to an end and asked the boys to get ready to eat dinner. Geoff seemed to want to continue the scuffle. At 5' 11" and 130 pounds, he was tall and slender for his age and in the throes of his adolescent physical and emotional development. Showing signs of pique at being forced to end his antics, Geoff insisted on starting a camp fire, and gathered the wood for it. Because of the snow, the wood was wet and wouldn't burn well. Geoff spent most of the pre-dinner hour nursing the fire by lying in the snow and blowing on the embers to get it to take hold. He got well-soaked with melting snow and again seemed irritated when he was required to change into dry clothing at bedtime.

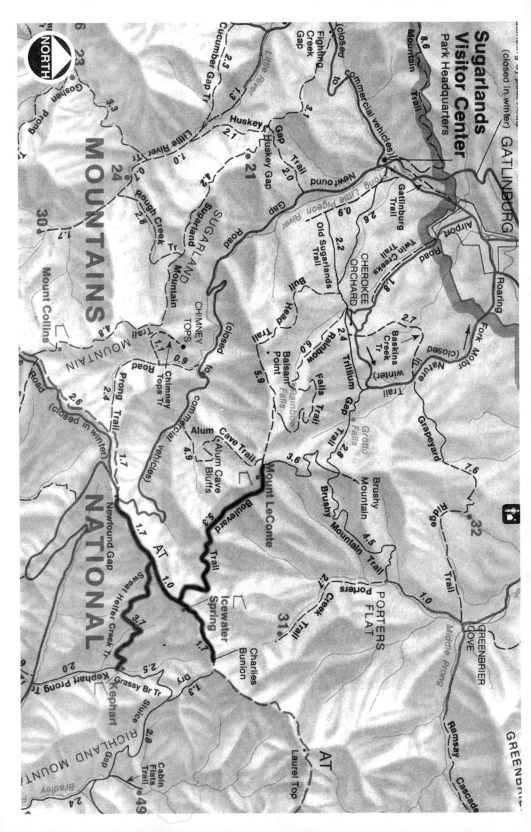

The next morning, Sunday, February 8th, the group woke to leaden skies, light snow and temperature below 20 degrees. They made breakfast, struck camp and packed up for the hike back to Newfound Gap. Geoff and Steve left the shelter with Scoutmaster Smith. Lee Smith stayed behind to secure the campsite, and was to follow the others out. Because he was the newest scout to have earned his explorer rating, Geoff was assigned to the middle of the hiking order with experienced hikers ahead of and behind him to be sure that everyone made it out safely. Lee was about twenty minutes in securing the camp site before he took to the trail.

Geoff was lagging behind as the other hikers in his group threaded their way through the woods on the trail. The path was well worn and distinct even in the light snow, but Scoutmaster Smith urged him to move along to stay with the others. When they reached the junction of the AT with the Boulevard Trail (BT), only about a half mile from their campsite, they stopped briefly to rest. It was now about 9:30 A.M. As the others again started to move out toward Newfound Gap, Geoff stood by himself, kicking at the snow. When Steve urged him to follow them, he hesitated and then said he would wait for Lee and come out with him. They left him standing at the junction near the trail signs which clearly point the way toward the BT and Mt. LeConte in one direction, the AT to Icewater Springs and Charlies Bunion eastward, and Newfound Gap, their destination, westward.

Scoutmaster Smith and Steve Wolfe reached the Newfound Gap parking lot and sat down to wait for the other two. Lee walked up in the swirling snow flurries, but Geoff was not with him. When asked about him, Lee said he hadn't seen him and didn't know he was not with the group.

All three hikers immediately retraced their route all the way back to Icewater but found no sign of Geoff. They then continued on another mile or so, as far as Charlies Bunion, thinking he might have gone back to meet Lee and somehow missed the location of the shelter at Icewater. Still no sign of Geoff. The group then retraced their steps back to the BT intersection and hiked out some distance in hopes of finding signs of Geoff in that direction. Again no trace. The light snowfall had been steadily increasing in intensity and by this time was covering any tracks or other signs of Geoff's passage.

Despite the snow, the hikers were puzzled by Geoff's disappearance because the trails were so well marked and the route back to Newfound Gap clearly indicated with signs pointing the way. It was now apparent that Geoff was nowhere to be seen. At the Point Last Seen (PLS), the BT intersection, he had been wearing a green corduroy coat, bright orange knitted toboggan hat and mittens, and was carrying his knapsack with food, extra clothing, matches and his sleeping bag. The rest of the group, now seriously concerned for Geoff's safety, quickly hiked back out to their car, drove down to park headquarters visitor center at Sugarlands and reported Geoff missing. It was now 3:20 P.M.

The report was relayed to the park dispatcher who alerted the chief ranger and others. District Ranger Mathis immediately dispatched rangers to run the trails leading to the Appalachian-Boulevard Trails junction. These search team members drove quickly to Newfound Gap, Cherokee Orchard, and Alum Cave trailhead areas and ran the trails converging on the AT-BT Junction. None of the first search team members was able to report any sign of Geoff. Because of the failing light, the increasing intensity of the falling snow, and the already heavy accumulation of snow further east in the park, no other searching was attempted this day.

Heavy snowfall continued through the night and by daylight on the morning of the ninth, another six inches had accumulated at Newfound Gap with deeper drifts in many places.

THE JOURNAL

Monday, February 9, 1970 6:00 A.M. - Day 2

Search Teams notified late last night are converging on Sugarlands at the Chief Ranger's office. They are outfitted for a full winter search. The plan, developed during the night, is for a trail search from all directions again toward the PLS at the Appalachian Trail (AT) - Boulevard Trail (BT) junction. Two rangers will come in from Pecks Corner in the east. Another team including Scoutmaster Smith will again hike in from Newfound Gap through the AT-BT junction to Mt. LeConte and on down to Trillium Gap and ending at Cherokee Orchard. Three others will bike up Alum Cave Trail to LeConte and return via the Bullhead Trail to Cherokee Orchard. Another group of three will also start at the Alum Cave trailhead, bike up that trail to LeConte and then down again via Rainbow Falls Trail to Cherokee Orchard. A fifth team of three will start at the hiking club cabin, search through Trillium Gap and end at Cherokee Orchard. A sixth will bike the Kephart Trail to Charlies Bunion. The last team will take the AT to the Sweat Heifer junction, move down that trail to Newfound Gap Road at the bridge. This will cover all of the trails which converge on Icewater where he might possibly have wandered.

Snow continues all morning. Despite the difficult walking conditions due to heavy accumulation of snow during the night on top of that which had fallen already, all of the planned trail sweeps are concluded without finding Geoff or any sign of him. During the day, arrangements are made with the Eastern Air Search and Rescue Service at Warner Robbins Air Force Base in Georgia and the Civil Air Patrol to participate in the search. Air search teams are standing by waiting for a break in the weather to join those on the ground looking for the lost scout. By early afternoon, arrangements are also made with the Sevier County Rescue Squad and the U. S. Forest Service for bloodhound dog search teams for assistance. Additional search personnel from volunteer groups such as the Tennessee Rescue Squad Association were solicited to join the search as well. Altogether about 41 searchers, heavily garbed against the snow, wind, and low temperatures are struggling up and down the dim trails calling and looking for Geoff all day.

Arrangements are made for articles of Geoff's clothing to be brought by his family so the bloodhounds can learn his scent. By late in the afternoon, trials with the dogs have been carried out on the AT where Geoff was known to have been, without any success. The dogs could not pick up the scent in the heavy wind and snow and very cold temperatures. All of these work against ground-tracking dogs.

The Sevier County Volunteer Rescue Squad set up their kitchen at Park Headquarters and will provide food for most of the searchers. They can handle up to 150 people each day both on site and off.

Late in the afternoon, the search management team, family members, Geoff's hiking companions, and volunteer search group representatives meet to discuss results and to plan for the next day. Geoff has with him the means to survive alone in the mountains, but the extremely bad weather and the prospect that he may have been somehow injured continues to increase the anxiety level of the searchers as the risks become clearer and time passes without a clue to his whereabouts.

The lack of success of the initial trail searches by the scouts, the hasty-search team, the larger group of 7 teams this day and random passage of other hikers who had been in the area and who were interviewed by rangers this day all force the conclusion that Geoff is not on the trails. He must have somebody gotten into an off-trail situation and is truly lost and unable to return to the trail system to be found. The plan for Tuesday includes searches of off-trail locations close to the PLS. This plan is understood to be a very difficult one to carry out because of the deep snow, high wind, and low temperatures. Active search operations in the field are shut down by 6:00 P.M. when darkness falls. No sign of Geoff. The snow is ending, but the wind is increasing and the temperature is dropping.

Tuesday, February 10, 1970 6:00 A.M. - Day 3

There are now 16 inches of snow on the ground with a low temperature of 14 degrees and winds of 25 mph. There are 133 searchers available to look for Geoff today. Included are rescue squadsmen from Blount, Knox, Sevier, Anderson and Hancock Counties; Morristown, Newport and Greenville, Tennessee; members of the University of Tennessee Army ROTC, the Carolina Mountain Club, the Smoky Mountains Hiking Club, the Cherokee NC Rescue Squad, a group from Clemson College, and numerous individual volunteers. They are here to search today in weather which is so inhospitable as to be extremely dangerous to any searchers who might themselves become lost or injured.

The logistics of moving these searchers and the food and equipment they need through the deep snow is a big challenge. First aid
and supply stations need to be set up at various central places like
LeConte Lodge and Icewater Springs shelter. Air support including
helicopters is available, but flying conditions prevent using them for
now. Over-snow transport vehicles are volunteered by an agency
and individuals, but these prove to be ineffective in the deep snow
on steep, wooded slopes. When loaded, they cannot move even
on the trail system until the snow is packed down. It isn't until mid-
afternoon that the helicopters are able to assist in transport and
search.

By nightfall, the results of the limited off-trail searching possible
today are negative; there is again no sign found of Geoff. Many
more searchers will be necessary to begin to search the off-trail
drainage areas surrounding the PLS. Plans for tomorrow at day-
break will be developed this evening; searchers instructed, and
team assignments made.

Wednesday, February 11, 1970 7:00 A.M. - Day 4

This is the third full search day. We try to use helicopters for
transport, but are unsuccessful due to high and gusty wind conditions at the hotspot which had been prepared at Icewater Springs
shelter area. If the helicopters can operate at all, it will be to the
parking lot at Newfound Gap which is also accessible to cars and
trucks. Moving searchers to the drainage basin search areas on
Porter Creek, Walker Camp Prong and the other basins which ring
the PLS will be done by over-snow vehicles when possible and, if not,
by hiking.

The helicopters are more effective as direct search observation
points. The view through the trees with snow on the ground makes
visibility good. Observers are able to see tracks in the snow and even
small animals moving about. When the wind dies, conditions are
promising for their use especially in the headwaters of the drainage
basins which are the most difficult and time-consuming to cover
on foot.

There are also plans to utilize dogs again. This time air-scent
dogs will be brought in to search areas before massive invasion by

searchers on foot in hopes of detecting signs of Geoff where he might have brushed against trees or shrubbery as he wandered along.

Efforts to reach areas in the eastern side of the park, beyond Icewater Springs, by snowmobile are unsuccessful due to the steep terrain and deep snow. These areas are being searched along the trails by air when conditions permit.

The highest temperature reached only briefly today during the afternoon was 38 degrees. Most of the day was below freezing and winds continued high and gusty.

A total of 289 searchers looked for Geoff today between dawn and dusk without any affirmative results.

Thursday, February 12, 1970 6:30 A.M. - Day 5

The temperature last night dropped down below 13 degrees but the wind has dropped considerably. Four air-scent dogs, German shepherds, and their handlers have arrived and are en route to the search area ahead of the other search teams. These dogs pick up scent from the air which originates from trees, shrubs, rocks, even water which has been in contact with clothing, skin, equipment, etc. belonging to the victim. Under ideal conditions, they can detect scent as long as a week after contact. However, these are less than ideal conditions.

Areas below the AT and Boulevard Trail are searched today, but it is very slow going.

There are again no positive signs of Geoff Hague today.

Friday, February 13, 1970 6:00 A.M. - Day 6

Weather is better today. It is clear, but low temperatures are still at about 12 degrees this morning. Dog teams are to start at LeConte and Icewater today working south and north on the Boulevard Trail. Scoutmaster Smith and a ranger are hiking to Pecks Corner and then on to Tricorner Knob in case Geoff somehow got beyond the original search ring and may now may be holed up at a shelter east along the AT. A plane with a public address system has been made available and is flying around the target search area calling for Geoff in case he is in some shelter and not visible. There is hope that *he will hear the calls and come out where he can be seen.*

The air-scent dogs have detected some scent, maybe of Geoff, along the Boulevard Trail about one and a half miles and again, three miles north of PLS toward LeConte. This tentative determination has sparked plans for more search dog teams and arrangements are made for 21 additional trail dog teams for tomorrow. They will come from Memphis, TN Police Department; Brushy Mountain State Prison, Petros, TN; Search and Rescue Dog Association, Renton, WA; US Army, Ft Gordon, GA and the Washington, DC Metropolitan Police Department. They will join the teams which have been working since Wednesday, including those furnished by Tom McGinn of The McGinn School for Dogs in Philadelphia, PA through Mrs. Billy Graham.

104 searchers have participated in the quest to find Geoff today to no avail except for the possibility of the air-scent alert along the Boulevard Trail.

Saturday, February 14, 1970 6:00 A.M. - Day 7

The weather is moderating some today. The low last night was 15, but the high yesterday got up to 40. Most of the day today should be above freezing. A light snowfall left less than half an inch of new snow but there is now more melting than accumulation even at the higher elevations. Wind is still strong however, hampering use of the helicopters. Low clouds and gusty winds reduce visibility even from the ground and the search is impeded accordingly. Some teams begin returning to their bases as early as 2:00 P.M. today. A total of 340 were involved in the ongoing search today.

Sunday, February 15, 1970 7:00 A.M. - Day 8

Weather is even milder today with a high yesterday of 37 degrees and a low last night of 30. Rain is forecast for today which should remove much of the snow cover now melting. This may be a mixed blessing since it will wash away tracking scent and make biting through wet mud as difficult as through heavy snow. It is tough going and most of those who have been hard-biting since the beginning of this search are showing signs of fatigue.

Search units are assigned to cover the trails and drainages leading from the Boulevard Trail between Mt. LeConte and the AT and along the AT from Icewater Springs shelter toward Newfound Gap. The Army dog team and the Washington State avalanche search dog team are assigned the primary search area along the Boulevard Trail where other dogs have previously alerted. This is in an effort to confirm and to determine if the dogs can actually follow the scent. The dog teams are to be sent in alone and regular search personnel excluded until they are done. The Memphis dog team will search the Porters Creek drainage, the Brushy Mt. team will search the Alum Cave area and the Washington, DC Metro unit will do North Carolina drainage along the Sweat Heifer Trail to just below the AT.

A report of finding a rock cairn and sticks pointed in a downstream direction on Roaring Fork near Grotto Falls was checked out but no evidence that it has anything to do with Geoff Hague. For the most part, tracks now reported in the area are proving to be those of searchers looking for Geoff.

The Army dog team reports that their dogs are getting definite alerts in the same general area along the Boulevard Trail as the previous hits by others. Since it is late in the day when this is confirmed, follow-up is planned for early tomorrow when the same dogs will have the chance to follow out the scent.

The evening planning meeting decides to scale down the operation and change from a search and rescue mode to a recovery mode. All volunteer search personnel except the dogs and handlers will be released tonight.

Three hundred fifty four people looked for Geoff today without success. Repeat of the dog alerts along the Boulevard Trail offer a small glimmer of hope. The real question, if they are indeed catching Geoff's scent, is whether or not he is still alive; it has been nine long, cold days since he disappeared.

Monday, February 16, 1970 6:00 A.M. - Day 9
The low temperature last night was 38 degrees with a high yesterday of 46. Light to moderate rain falling today and over a half

inch overnight, melting most of the rest of the snow. The forecast calls for rain to end today, but continuing windy at 15-20 mph; clearing tomorrow and Wednesday with winds diminishing.

Assignments today include the Washington State team on Anakeesta Ridge, the Army team on the drainage of Walker Camp Prong below the Boulevard Trail; the Metropolitan DC team over Mt. LeConte on the Rainbow Falls section and Bullhead; McGinn teams on the ridges in North Carolina between Newfound Gap and Sweat Heifer Trail; the Memphis team on Indian Gap Trail down to the Chimneys.

At about noon, searchers discover a pack with some items of clothing and equipment on a rock in the middle of Walker Camp Prong about a mile down stream from the Boulevard Trail toward the Newfound Gap Road from Sugarlands to the gap. These were brought in to Headquarters at approximately 2:45 P.M. and identified by his parents as belonging to Geoff. They were inventoried and it is concluded that wherever Geoff is, he does not now have his necessary supplies and, because none of the food has been consumed and everything else is intact, he appears not to have used his pack since he was last seen. Speculation has led to the idea that he may have put down the pack to find wood to build a fire and either become separated from it in the storm or even have fallen into one of the many pools along the stream and drowned. The same Army dogs and handlers who found the pack follow the scent downstream all the way to the Newfound Gap Road where it is lost. Plans are made to send searchers back into the same area including some of the same dogs and handlers who made the find. Only 40 searchers participated in looking for Geoff today.

Tuesday, February 17, 1970 6:30 A.M. - Day 10

Low temperature last night was 34 degrees with a high of 43 yesterday. It is foggy in some low-lying areas but otherwise clear.
There is continuing high interest from the media in this search. The first find, of Geoff's pack yesterday, has fueled all kinds of speculation and a press conference is held to brief the media. They are shown Geoff's belongings and asked not to enter the Walker Camp

Prong drainage, which is easily accessible from the Newfound Gap Road until the area has been thoroughly searched.

The search plan for today is to cover the Walker Prong Drainage and to perform a detailed search both up and downstream from that point to the Boulevard Trail and Newfound Gap Road until more evidence of Geoff is found. One team is assigned to search below the road along the Walker Camp Prong. The Army team returns to the location where the pack was found and again follows the scent downstream to the road. Their conclusion is that Geoff followed the stream down to the road and is now out of the search area and perhaps out of the park. The search by other dog teams below the road reveals no sign of Geoff's scent, thus reinforcing the opinion of the Army team. Based on this conclusion, the Army banders further decide that they can do no more to assist in the search and they propose to return to their base.

When this information is communicated to Geoff's parents, they say that they believe that if the dog team's conclusion is correct, their son must be suffering from some serious illness or loss of memory since they believe he would not normally run away, especially under such strange conditions. With the approval of his parents, a description and a picture will be circulated to the news media and law enforcement agencies to try to get some response about Geoff if he is outside the park. Media representatives are assembled and briefed, and plans made to continue searching in the park the next day with the remaining teams. Only 36 people searched for Geoff today.

Wednesday, February 18, 1970 6:00 A.M. - Day 11

Weather continues mild today with a low last night of 34 degrees and a high yesterday of 47. It is clear today with some valley fog. Helicopters are able to fly. The remaining personnel and dogs are assigned to continue the search in Walker Camp Prong. The Washington State team will follow the Walker Camp Prong all the way from the Boulevard Trail down to the place where the pack was found. They start at 5:40 A.M. to hike to the beginning of their search area so as to be able to take full advantage of the daylight.

The Green Beret team with Tom McGinn and his dogs are assigned to the slopes of Anakeesta Ridge on the right flank of the Washington State team. Rescue squad personnel are assigned to search remaining areas not yet combed in the vicinity of the PLS. NPS crew leaders with radios are assigned to these search teams to assist in communicating any information found to headquarters. Helicopters are to assist in transporting searchers in and out of their assigned areas.

At 10:17 A.M. a coded transmission from the Washington State team is received at Headquarters that the body of Geoff Hague has been found. The message is confirmed by other personnel who immediately move to the area to assist in transporting the body. After verification, the news is communicated to Geoff's family and ultimately to the media to call off the widespread search for further information.

Geoff has been found about 1000 yards below the BT, about a mile from PLS. He is in the Walker Camp Prong drainage about 1000 yards upstream from the location where his sleeping bag and pack were found two days ago. Geoff is at the base of a tree; half-sitting, half slumped into a fetal position and is still covered with deep snow except for part of his right arm and part of his right leg where the snow has melted off him. He has no socks on and one boot is lying 6 feet away. The other is unlaced and half on his right foot. His orange toboggan cap is off, his mittens are off, his coat is open and partially off. His left arm is pulled up inside his sleeve. His shirt is on but his pants are unzipped and partially removed.

In the snow which still covers the body and the area around it are dog tracks from previous days searches. About 12 to 15 feet away human footprints are visible on the same side of the stream and between Geoff and the water.

Geoff is placed in a litter and carried out to Icewater Springs Shelter by the same teams which searched for him so intensely for many days. They arrive at Icewater about 2:15 P.M. From there he is taken by helicopter to the Sevier County airport and then to the County Hospital for medical examination. County Coroner Harold Atchley and State Medical Examiner Dr. John Hickey both examine *the body. They conclude that although exhibiting minor scratches and bruises, the body does not have any broken bones or other serious injury. He is believed to have died of exhaustion and exposure late Sunday, February 8, or early Monday, February 9, 1970.*

What happened to Geoff Hague? How could he have disappeared so quickly without a trace and how and why did he wind up deep in the drainage of Walker Camp Prong, half undressed and without his pack and sleeping bag which could have saved his life? There are no clear or easy answers to these questions and since Geoff is not here to tell us, there may be no answers at all. However, there are several tantalizing clues and there may be some conclusions we may reasonably draw.

Based upon his apparent mood in the hours before his disappearance, Geoff may have been feeling restricted and wanting to express his independence from the group. It is a common occurrence for hikers on the AT near the Boulevard Trail or actually on it to hear vehicles ascending the Newfound Gap Road to the gap. Autos frequently sound their horns at the spiral underpass which is just where Walker Camp Prong crosses the highway. These horn sounds give the false impression of being quite near to the trail when they are really almost a mile away. It is possible that Geoff heard these auto sounds and thought he would take a "short-cut" to the parking area which is not too much different in direction from the source of the sounds. If he did indeed choose this way to exercise his adolescent independence it cost him his life, for the path thus chosen put him into an inhospitable environment under very difficult weather conditions and with little experience to protect him.

Once he started down the slope through rough brush laden with new snow, he was probably soon wet and beginning to experience symptoms of hypothermia. By the time he got to the stream he was probably very cold, wet and exhausted, and perhaps disoriented as well. Although he had come less than a half mile, under the conditions encountered it must have seemed much further. When he realized his mistake he probably turned to go upstream to try to regain the trail. His heavy pack would have seemed even heavier to the slender youth and at some point might have seemed too heavy to carry up slope and was abandoned in the stream. This was a second and even more tragic mistake, for without his matches, dry clothing and sleeping bag he stood not a chance of recovering his errors without paying with his life.

As the irrationality of hypothermia invaded his mind as a result of the drop in body temperature , he very likely imagined that his feet and hands were burning hot and removed his shoes, socks, and gloves, which hastened freezing, the inevitable coma, and eventually death. Based on the distance he was found from the trail, if this scenario is indeed reasonably accurate, he was probably dead within six to eight hours of leaving the trail; an extremely high price to pay for youthful rebellious feelings and inexperience.

SAVVY LADY • Cindy Lee Webster

On Friday afternoon, March 22, 1974, Cindy Lee Webster, a 26 year old outdoor education teacher from Bloomington, Indiana checked into the Greenbrier Ranger Station where she applied for and received camping permits for the 3-day solitaire hike she had planned. This trip was a scouting expedition to plan a special trip for later in the year which she hoped would challenge her students more than a typical hike on the well-used trails of the park. Especially interested in the growth and development of her charges, she wanted to take them on a cross-country hike using compass and maps and dead reckoning to find their way.

Her permits called for her to spend Saturday night, March 23rd, at Porters Flat on the headwaters of the Little Pigeon River above Greenbrier; the second night at the Icewater Springs shelter on the Appalachian Trail (AT), and the third night at the shelter on Mt. LeConte. She told the Greenbrier ranger that she planned to come back to the station via Rainbow Falls on Tuesday morning before heading to Indiana for her birthday party at her parent's home on Tuesday night.

On Saturday, she pitched her camp at the Porters Flat shelter where she visited briefly with another camper. The weather had been overcast all day but the forecast was for decreasing cloudiness on Sunday. Temperatures were cool, but generally above freezing at the lower elevations during daylight hours. On Sunday she woke to heavy overcast and, as she broke camp, she watched the leaden sky for signs of snow. She took a compass reading to establish a direction for her cross-country traverse to intersect the AT at the south end of Porters Mountain near Porters Gap just east of Charlies Bunion.

Cindy left the Porters Creek Trail at its mid-point at about 3000 feet elevation, headed southeast, and angled up the mountain in what at first was pretty easy going in open second-growth forest. However, as she climbed the undergrowth thickened and became

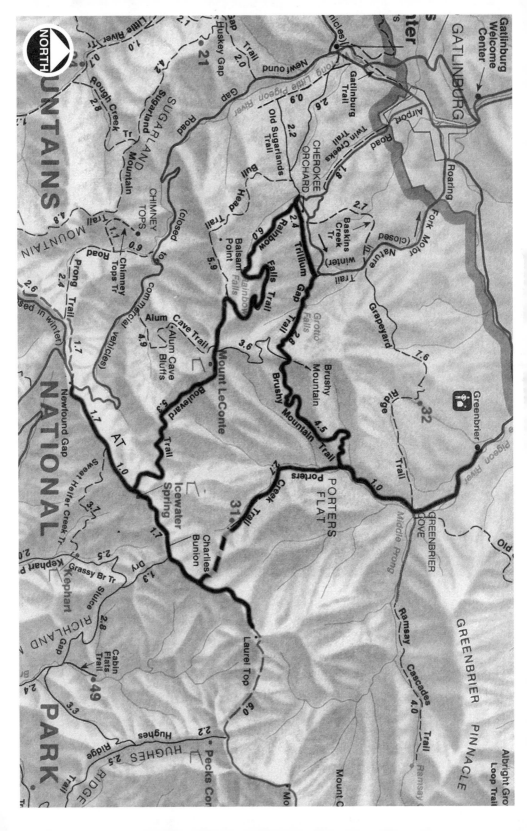

closed and unyielding. At the same time, contrary to clearing as predicted, the weather became even more densely overcast and it began to snow. The combination of dense undergrowth and falling snow made it difficult to see landmarks and to hew to the line she had set for the off-trail traverse. Soon she was lost in ash thickets, rhododendron and laurel "hells" on a steep slope which made it difficult even to keep her footing in the accumulating snow. Finally, she was reduced to crawling on her hands and knees, and was getting wetter and colder by the minute. That's when she lost her map.

From her earlier experience in wilderness hiking and two stints in outdoor survival schools, she knew the perils of exhaustion, hypothermia and disorientation. She already felt the effects of the latter and began seriously to fear the others. She also realized that if she fell or otherwise injured herself under these conditions that she would be unlikely to survive. These thoughts filled her mind as she struggled up the slope inch by inch trying to reach the crest of the ridge without further risk.

THE JOURNAL

Wednesday, March 27, 1974 6:30 A.M. - Day 4

When Cindy Webster failed to appear for her birthday celebration at her parent's home in Indiana, her mother reported her missing to park authorities late Tuesday evening, March 26th.

The call came very early this morning and I am requested to report to the Greenbrier station for assignment to the search. There are 75 searchers including park personnel, members of the Sevier County, Blount County, Morristown, and Newport rescue squads, plus students from Lake City Junior College who have been participating in a week-long outdoor laboratory study in the park, all gathered here to look for Cindy Webster. The weather is cloudy and cold with temperatures in the search area in the 20's and low 30's in daylight and in the 15 to 20 range at night.

After a quick briefing, I am assigned to a hasty-search detail, one of several searchers running trails leading off from the Porters

Flat campsite where a hiker who observed Webster there on Saturday, March 23rd reported their last contact. All trails are searched on her intended route with no one observing her beyond Porters Flat. No sign or clue indicating Webster had traveled them was found on these trails. The helicopter is not able to fly most of the day due to heavy cloud cover, wind and some snow. Webster is not located this day despite intensive searching. The conclusion is that she is not on the trail system and must be somewhere off-trail.

Thursday, March 28, 1974 6:00 A.M. - Day 5

It is still overcast today with temperatures hovering around freezing.

Today the search will be concentrated around Point Last Seen, Porters Flat campsite not far from the Greenbrier Hiking Club building. I am assigned to track Cindy if I can, beginning at the Porters Flat Shelter. Since my first search for Dennis Martin in 1969, I have had several opportunities to work with tracker JR and others who have learned to read the woods and I am anxious to try my own hand at this interesting and useful art.

After casting about, I am able to locate a dim trail of one person exiting the Porters Trail and climbing east up the mountain. The trail is dim and, I can tell, some days old. As I track the person, from time to time I lose the sign but fortunately find it again as I move up the mountain. I come through an old location where a board cabin had once stood long ago. There are water pipes leading from a spring and a leveled area where piles of rotting boards are lying. I have no idea what this is but there is no sign that the trail I am following is in any way related, so I continue on following. The terrain turns very steep and rugged with very dense undergrowth, and large first-growth trees. As I follow the tracks, the slope becomes even steeper and the climber ahead of me begins to sidehill and switchback leaving deep footprints in the snow and dark soil. The roar of Porters Creek has long since left me and the forest is quiet except for the sound of the wind in the trees and the crunch of my boots in half-frozen soil. An active breeze is blowing upslope from the creek and across the ridge I am climbing.

Despite the heat generated from the climb, the wind is cold at the higher elevations and I sometimes feel a chill and begin to shiver.

Wild onions appear here growing in great quantity on this slope and are just popping through the forest litter. It occurs to me that I might be following a person who was digging these wild onions, called "ramps", earlier in the week. At lunch I pause to consume some of these onions myself, initially to my great delight but with later regret as I pay for my indulgence with much gas and bad breath from this garlicky wild onion.

Later, I notice a helicopter make a pass over me as the clouds are trying to clear. Continuing to follow the dim trail through brush so thick I can see only short distances, it is obvious now that this is no ramp digger as the tracks I am following are well above the ramps. As I climb, the tracks pass an area to the left which appears very unusual. Some person a long, long, time ago had been to this area repeatedly. There many signs of human disturbance of the terrain and evidence of old excavation in a rock outcropping that is now covered with many years of debris. I do not understand this area and continue up the mountain as it appears to have no relationship to the search.

Farther up Porters Mountain, the slope flattens as I near the crest and the footprints become less distinct in the shallower, more rocky soil. The top part of Porters Mountain is an almost continuous sea of thick laurel forming a brushy "bald." I can see where my target climber, whom I believe to be Webster, has pushed through the "hell" thicket by the branches broken in the direction of her movement. I do the same thing following her and comparing the appearance of the broken twigs. This was not done by an animal, even a bear, which would have passed with far less disturbance. My broken twigs are green and fresh and hers are already brown so I can conclude that my quarry is perhaps 2 to 3 days ahead of me. Passage through the laurel is difficult and slow and I can see that my predecessor has stopped often to rest. Here and there a deer trail and tracks of fox and a bobcat are encountered as I push on through the thicket. Overhead, I see a red-tailed hawk circling on the updrafts.

Suddenly the helicopter again makes a pass over me in the direction which has been established based upon my tracking direction and begins circling the top of the ridge above. Over the radio comes word that Webster is spotted on top of Porter Mountain in a thick laurel patch. A glance at my watch says the time is approximately 4:50 P.M.. I feel a thrill that Webster is alive and active enough to be waving to the chopper as I abandon her trail and proceed to climb directly toward the circling helicopter. Food, water, and survival gear are dropped to Webster and I hear the thudding sound of the chopper blades more distinctly as I near the top of the ridge.

Once on top, it is still necessary to bulldoze through more thick laurel to reach the chopper location and Cindy. My radio is on and the volume turned up so I can hear the search traffic as I approach her camp. She cannot see me coming through the dense laurel but can hear only the breaking of brush and the static of the radio which frightens her. As I near her location, she begins to scream loudly. To try to calm her fears, I stop and call to her that I am a searcher who is looking for her.

Entering the camp she runs to me, hugs me and clutches my arm and talks my ear off. She says she has been out of food for several days and has melted snow for water. I give her my water and some food I have. After awhile other searchers come up the mountain and we gather up her equipment and start off down the mountain and again this time at a slow pace in the gathering darkness.

Webster is O.K. except for dehydration. She says she has been at this location for the past five days conserving her small supply of water. The truth of this is self-evident in that she is a little odorous but alas, because of my ramp feast and the sweaty climb up the mountain, I smell no better. Taking our time we walk her out to the trailhead, arriving at midnight.

Webster is a nice lady, friendly, and pleasant to be around. As we hike out, she tells me her story. She says her original intention was to travel cross-country from Porter flat to the Appalachian Trail via Porter Mountain. However, when she ran into trouble, she decided right away to make a camp at the first opportunity and *wait until conditions improved or she was found and rescued. She knew that if she was delayed enough not to make it home for her birthday, her mother would report her missing and the park authorities would launch a search. This smart decision may have saved her life.*

When she reached the top of the ridge after her arduous and exhausting climb, she found a suitable spot, cleared some brush and made camp. Mindful of the need to be visible from the air, she pitched her small tent in plain sight in her little clearing. Later she hung a spare parka and her knapsack in the nearby trees to further attract attention to her location. For then, she was content to eat some food and hole up in her little tent to warm up and wait for daylight and the snow to end.

The next day, Monday, she woke to a winter wonderland of snow surrounding her tiny campsite. Several inches of snow had fallen and she knew that the combination of snow and difficult terrain would immobilize and isolate her for days. She gathered wood for her fire as best she could in the snow, melted some snow for water to supplement her dwindling canteen supply and generally set about making herself comfortable. In this way she spent not only that day but three more before the helicopter spotted her. During that time, as her small food supply rapidly diminished, her anxiety level increased.

When we reach the Greenbrier Hiking Club Building at the trailhead at midnight, Cindy is driven out to Sevier County Hospital where she is examined and pronounced physically O.K. except for dehydration.

On March 29th, I return to my regular duties, thinking about how our search techniques have changed over the last five years. We now routinely conduct hasty searches first, followed by tracking visually and with the aid of dogs. If weather conditions are with us, these methods seem to be more effective than the massive grid searches sometimes employed.

Even today, my thoughts sometimes return to the mystery surrounding the signs of human activities which I observed high on Porter Mountain that spring day. What was it? I muse, reflecting that the Smokies have many such lost places. However, in a March 8, 1988 issue of the Knoxville Journal, I am surprised to read of the legend of Perry Shults's old silver mine that he is said to have located in 1867 on the slopes of Porter Mountain. It is said that Shults had found a rich vein of gold and silver ore and over many years came secretly to visit the site, leaving his wife setting on a rock near Porter Creek at the base of the mountain. My mind flashes back to the disturbance I had observed on the Mountain that day in March of 1974 some fourteen years prior. I had almost forgotten the find. What was it, I asked myself? Could I find it again? Was it his mine, or was it where relatives of Shults had explored and dug in the mountain in hopes of finding his secret? Maybe it was neither. As most legends of lost mines relate, Shults evidently died without taking his family to the location. I never went in search of the spot I saw again since I know that mining claims are not permitted on national park lands and finding the site might only prove a problem to the park service. I did not document in any way my route up the mountain that day and the observation remains only a notation in my NPS issue pocket notebook.

PRUDENT PEERS • ERIC JOHNSON & RANDY LAWS

While Eric Johnson and Randy Laws were earning their merit badges for outdoor crafts, the two fifteen year old eagle scouts had hiked and camped most of the Appalachian Trail in the Tennessee-North Carolina area. Thanksgiving weekend in 1974 was a school holiday which they decided to spend in Great Smoky Mountains National Park. They planned a backpacking trip from Davenport Gap at the extreme eastern edge of the park where the Appalachian Trail (AT) enters to Newfound Gap. After checking in at Sugarlands in Gatlinburg for their camping permit on Friday, November 29th, Eric's parents dropped the boys at Davenport Gap and headed back to Johnson City, Tennessee, planning to pick them up at Newfound Gap on Sunday at about 6:00 P.M. The boys were well prepared for a late fall hike with thermal socks and underwear, warm sleeping bags, and extra food in their packs. Because Eric and Randy were experienced hikers and campers who had made careful preparation for this expedition, Eric's parents had few qualms as they drove away from the trailhead.

Shouldering their packs, Eric and Randy started the long climb westward on the AT. Although it was overcast and near freezing, the trail was clear and the two hikers had no problem moving along at a comfortable pace. Near noon they stopped for a rest and some lunch at Low Gap, noticing that the sky was lower and darker with scudding gray clouds. By mid-afternoon and while rounding Mt. Guyot they felt the first flakes of falling snow. By the time they reached the shelter at Tricorner Knob, where they intended to spend the night, the wind had picked up and snow was falling heavily. They gathered a good supply of deadfall wood for the fire, made an evening meal from their supply of freeze-dried foods and some fresh fruit and cookies, and watched the falling snow for awhile before stretching out in their sleeping bags for the night.

The next morning snow was still falling and, whipped by strong winds, accumulated in drifts as deep as five feet with an

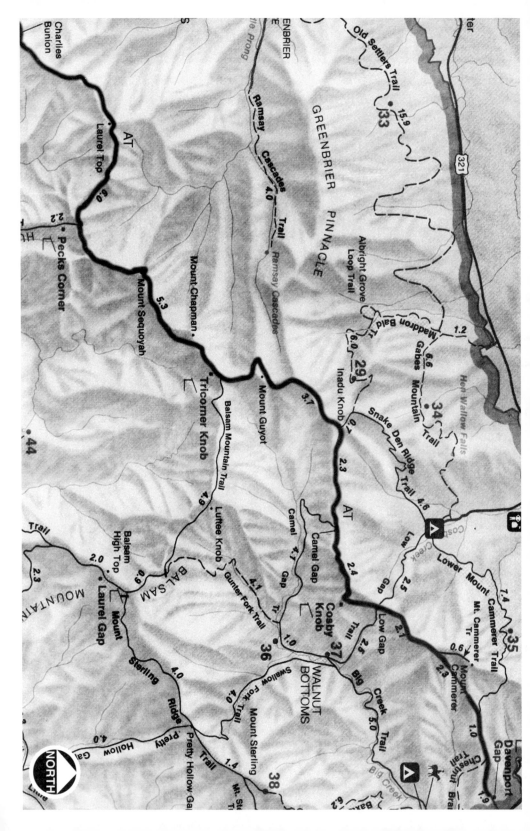

average depth of over 18 inches by nightfall on Saturday. As the boys lace up their boots and prepare to leave the shelter, they do not realize that it will not be until Sunday afternoon before the early winter storm will blow itself out with between two and three feet of snow on the ground and strong winds piling it chest-high everywhere in the mountains.

On Sunday night, Eric's parents, concerned about the weather generally, are shocked to find that access to Newfound Gap via the Newfound Gap road is cut off by the heavy snowfall and the road is closed. They cannot make it to the pick-up point for the boys and they begin to have real fears for their safety. Enlisting the aid of the park rangers, they are taken by four-wheel drive vehicle to Newfound Gap, but find no sign of their son and Randy Laws by the appointed 6:00 P.M. rendezvous. They wait until it is quite dark and they are certain that the boys are not nearby, then return to park headquarters to wait while a search plan is developed.

THE JOURNAL

Sunday, December 1, 1974 10:00 P.M. - Day 2

Word comes via the telephone from Assistant Chief Ranger Jack Lanahan about the missing scouts and the search planned for tomorrow morning. My thoughts go immediately to Geoff Hague and the tragic end to that search and, as I ready my gear for the next day I try to be optimistic about this one. By the time everything is ready, it is late. It's a short night before the search begins.

Monday, December 2, 1974 7:00 A.M. - Day 3

When I arrive at Newfound Gap via jeep from park headquarters, it is still dark. A team of searchers with a snowmobile from Cosby has already been dispatched up the Snake Den Trail to the AT. Another is also already on its way east from Newfound Gap. A third, on snow shoes, is attempting to walk in along the same trail toward Icewater. I am on standby at Newfound Gap trying to keep warm and awaiting active assignment. Radio reports from the search teams already under way indicate very slow progress. The wind is strong, buffeting personnel, trees and everything else with sharp gusts carrying stinging snow. Visibility is very low. We learn that the team from Cosby has gotten off the trail into deeper snow and is bogged down and cannot make further progress. They are fighting now to extricate the machine and make it to a shelter.

The weather is so bad that air search personnel cannot fly because of poor visibility and high wind. Ground search is very slow, hampered by deep snow and the biting wind which saps energy and induces hypothermia in minutes. I am frankly happy to be relatively warm and sheltered for now, but hopeful that conditions will improve and we can expand the search. Like other "organized" activities, searches seem to be composed of equal parts of strenuous activity and waiting. Right now I am waiting, but I know that soon enough I will be sweating and freezing at the same time when I am assigned actively to the search.

By mid-afternoon, that moment comes when I am assigned to hike in to Icewater Springs shelter where the snowshoe team has holed up exhausted from their trek, having found no sign of the boys. It has taken them nine hours to battle the three miles to Icewater, and they are completely spent. The snowmobile team from Newfound has also had continuous problems battling the deep snow despite their machine assistance and is returning to Newfound while we assist the snowshoers.

We hook up with the crew at Icewater and help them battle back toward Newfound Gap. By nightfall we are all safely at the parking lot and after a short rest to warm up, we return to headquarters, and then home for the night.

I have never seen conditions as bad as these and I am very apprehensive about the young men we are trying to find. If the full brunt of the storm caught them away from shelter, I know that they cannot survive very long under these conditions.

Tuesday, December 3, 1974 6:30 A.M. - Day 4

This day dawns bright and clear and, as I prepare to join my companions for the search, I am hopeful that air search can be used today because it may be the only way we can find the boys before it is too late. As I approach Newfound Gap in the jeep, I can hear the helicopter and soon see the Vertiflite chopper from the University of Tennessee with "Doc" Lash and his capable assistants. They will fly the trail east checking both the trail itself and the shelters between here and Davenport Gap to see if they can spot the scouts or any sign of them. The rest of us will try again on foot but we know that it will take many days to search even the trail system in this deep snow. There is no hope of covering any area off-trail if they are not found directly on the trail system or shelters.

*Within an hour we hear from the radio that the helicopter has spotted the words **H E L P** stamped in the deep snow at Tricorner Knob Shelter and that there are signs of habitation there. A red packsack is visible next to the shelter and soon two people are reported to be seen jumping up and down in the snow and waving to the occupants of the chopper. It looks like these are the lost scouts and we are all also jumping and shouting with joy. The wind is still too strong and gusty for the little UT helicopter so a message is relayed to Fort Campbell, Kentucky where a rescue unit is already on alert. By early afternoon a big Chinook is on its way. In the meantime, "Doc" Lash and his crew have dropped emergency supplies to the boys. Word has been relayed to Randy Law's parents by the Johnsons that the boys have been found. We do not yet know their condition, but the fact that they are able to move around to signal their rescuers is a very good sign.*

By 3:30 P.M. the Chinook is in position and rescue personnel are lowered to the ground. They harness the boys and help them into the winch equipment which raises them into the chopper where the medical team gives them a quick exam on the way to the hospital at University of Tennessee. The boys are apparently O.K. except for some frostbite on Eric's toes. Parents are on their way to the hospital and we begin demobilizing the search operation and preparing to go home.

Everyone involved in this search was happy that the scouts had had the unusually good judgment to remain in their shelter instead of trying to rescue themselves. They told their parents and rescuers that they tried to hike out on Saturday but found the snow so deep that it took more than two hours to go less than a quarter mile. Wisely, they knew that their best chance of survival was to conserve their energy by returning to the shelter, finding firewood to keep warm, and rationing their food until they could be found. Other than a little dehydration and minor frostbite suffered by Eric, they survived their ordeal quite well.

Eric Johnson finished high school and college and went on to law school but lost track of Randy after their high school days. Eric continued his interest in hiking and has enjoyed many outdoor adventures, having climbed several challenging mountains including Mt. Rainier in Washington and Mt. Mc Kinley in Alaska.

SPRING BREAK • MARK HANSON

March is entering like a lion. It is terrible, terrible weather; cold and cloudy, with high winds of 40 mph at noon this March 7, 1975. The clouds are so dark visitors on the access roads to the park must turn their headlights on to see. Three tornadoes hit Cades Cove destroying 1000 trees in Russell Field and Cable Mill Area. Live trees are sheared off half way up the trunk. Hail falls, stripping dead branches from the trees, and accumulating in some areas as much as four inches deep. The weather continues highly variable during the next two days, bringing flooding in the low lying areas and extreme cold and heavy snow driven horizontally by high winds. It drifts up to five feet in higher elevations. Ridge lines and the trees which soften their razor edges are virtually invisible.

Entering this scene in the relatively calmer and sheltered conditions at lower elevation are three backpacking friends off on a bold adventure. Mark Hanson, Ben Fish, and John Chidester, all students from Eastern Kentucky University on a spring break from school. They are all in their early twenties. Despite the threatening weather, they park their car and start their hike from Big Creek near Mount Sterling at the extreme eastern edge of the park. They are all wearing typical student garb of blue jeans, sweatshirts or sweaters, jackets and hats. They plan to travel west toward Tricorner Shelter where they expect to spend the first night.

Starting at an elevation of about 1700 feet, their trek will take them almost continually uphill to about 6305 feet. The hiking distance will be a little over 16 miles to Tricorner. Climbing Sunup Knob about six miles from their start and at an elevation of about 4500 feet, they begin to encounter increasingly sharp wind and heavy snow. After continuing about two more miles directly into the biting wind, and near Low Gap, John Chidester complains his boots are wet and cold and are not doing the job. He decides to turn back to the car and leaves the others. Soon after, and less than a mile beyond the place where Chidester left them, the remaining two hikers pass the shelter at Cosby Knob but they do not attempt

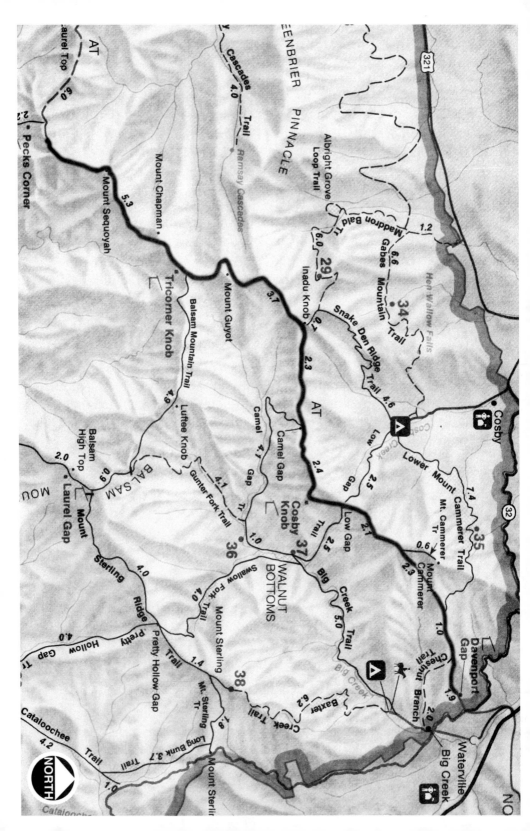

to stop and camp here. They are still less than half way to their intended campsite.

Hanson and Fish continue uphill full-face into the bitter wind. By the time they reach Camel Gap in midafternoon, Hanson is exhausted and already well into the throes of hypothermia. He tells Fish that his pack is too much of a burden and drops it on the trail. Fish tries to convince him that he needs his sleeping bag and other essentials but Hanson mistakenly says that the shelter at Tricorner will have whatever he needs. He abandons the pack. They continue on, eventually reaching Mt. Guyot. By this time it is around 7:00 P.M. and darkness is falling rapidly as it does in the mountains. Fish still has his pack and trudges on in the lead in the semi-darkness. The storm is unrelenting in its intensity and, without the psychological benefit of daylight, seems even stronger and colder as night falls.

At about 8:30 P.M., there is total darkness. They are now only 400 yards from Tricorner Shelter when Hanson, exhausted, gives up. He sits down on the trail and refuses to go any further. Fish tells him he will go for help . He does not know the shelter is only four hundred yards away. After going only two hundred yards, he stops and also gives up. Somehow, he manages to pull out his sleeping bag and crawls into it during the night. Completely spent and numb with cold, he dreams he hears cries for help. It was probably no dream as Mark Hanson was only a short distance away.

When Fish awakens at about 7:00 A.M., he searches for Hanson but cannot find him. Everything looks different in the daylight and is covered with a thick blanket of new-fallen snow. After spending some time wading through drifts on the trail and calling for Hanson, Fish continues on to Tricorner Shelter where he finds two other backpackers camped. They return with him to where Hanson was last seen and search the vicinity but cannot locate him either. They all hike out to Cosby and report Hanson lost at the ranger station there late in the afternoon of Monday, March 10th. It is too late to mount a formal search, but Fish returns with Ranger Acree on an All Terrain Vehicle to verify the exact location where Hanson was last seen.

THE JOURNAL

Tuesday, March 11, 1975 7:00 A.M. - Day 3

I am among several searchers responding to Tricorner from Cosby. We hike up Low Gap Trail and turn right toward Tricorner on the AT. It is extremely cold with heavy snow. I observe as we walk up the trail that where the sun shines now and then on the south-facing trail, there are patches of muddy wet on top of the frozen earth, but as we move over to the north facing side of the ridge it is still frozen solid and icy with deep snow off the trail. Animals like bobcat and fox are active throughout the spruce fir forest and red crossbills dart here and there. Snowbirds scratch in the frozen snow for seeds. Red squirrels scold us as we proceed up the trail. The cold is penetrating and the snow-laden wind blows fiercely out of Tennessee through gaps in the ridges. The wind-driven snow crys-tals are blown horizontally by the force of the wind and they sting as they strike the few uncovered patches of skin on my face. My feet are dry though, because I am wearing heavy Sherpa boots and my legs are warm in my heavy woolen pants which do not freeze.

On the trail at Camel Gap we find Hanson's pack and equip-ment and slow our pace in order to look for him more carefully here, but there is no sign of him. We go on, and as we near the junction of Snake Den Ridge Trail with the Appalachian Trail, I divert to an old airplane crash 100 feet off the trail with the thought that Hanson may have taken shelter in it. This crash occurred in November of 1962 and killed two military officers. One of the offi-cers lived a while and was able to contact a ham operator. His last words were, "Crashed somewhere in Smokies, badly injured, cannot last long, may God have mercy on our souls."

Well, I could not find where anyone had been around the plane so we continued on toward Tricorner shelter, arriving there late that evening. Since other searchers have already filled the shelter, I stay in a tent up on the hill above the structure. We did not locate Hanson or any sign of him this day.

Wednesday, March 12, 1975 6:00 A.M. - Day 4

Since there is no sign of Hanson on the trails, the next step is to carry out an off-trail search. We are split up in ten small search groups and my group is assigned to search in a loose grid pattern from PLS west toward Mount Chapman on the Tennessee side. We spread out down the mountain and begin our sweep. After going about a quarter mile, someone uphill of me finds an old cave in the side of the mountain. It needs to be looked at inside but the other searchers are hesitant. I agree to search the small cave, but aware of the possibility that there may be a sleeping bear inside, I decide to back into the cave butt first. My reasoning is that I can afford to lose a little butt but I can't afford to lose my head. I find indications that either a bear, boar, or large cat has used the cave numerous times. I find a type of fungi called Devil's Snuffbox growing on the floor and there are very old bones scattered about. I do not find any indication Hanson has been here so I exit the cave and continue the grid search west of Tricorner.

The forest has been fairly open so far, but as we near Mount Chapman it gets thicker with downed trees, briars, and laurel. As we swing around the western flank of Chapman our search comes very close to another plane crash, this one of an old Cessna 182 that came down on February 2, 1969 but wasn't found until May 31, 1970. Bears had consumed the body of the pilot, John C. Koppert, the only person aboard. I determine Hanson has not taken shelter here either and so we continue our grid back to Tricorner.

We continue west to Mt. Sequoyah, searching the Tennessee side of it, then shifting to the North Carolina side, and circling east back toward Tricorner. The terrain is more gentle than the Tennessee side and I find Mt. Sequoyah very pleasant despite the heavy rains that are falling on us now. The temperature has warmed to 56 and the snow is melting. The woods are wet and muddy and it's impossible to stay dry. Proceeding East toward Tricorner we come upon two more old plane crashes that occurred within 5 days of each other. The crashes are both less than 1200 feet from the Appalachian Trail and are only 400 feet apart. The first was an army two motor C-45 plane with six persons aboard. It crashed on

10/12/45. All persons including a WAVE were killed . The second plane was a two person L5 Cub sent to photograph the C-45 on 10/17/45. It stalled into the trees and neither the pilot nor the photographer were hurt. They walked out on their own.

There is no sign of Hanson at these two locations either and we proceed with the search. We encounter heavy plowdown on the North Carolina side of Chapman and it slows our progress. Beyond this area the forest is more pleasant. We proceed to the Old Horse Camp located about a quarter mile west of Tricorner on the North Carolina side. It is only 200 feet down to the site but we find no evidence of Hanson having been there either.

Darkness is approaching as we arrive back at Tricorner. It seems there is now a vacancy available in the shelter and I quickly fill it. It is quite comfortable inside as plastic has been strung across the front to block out wind, rain, and snow. I have one of the lower bunks and I wonder why it is vacant. After a hurried meal, I turn in and the reason for my good fortune becomes apparent. It turns out that my bunk happens to be located directly beneath one occupied by two amorous searchers who keep me, and I suspect many of the others in the shelter, awake with their nocturnal activities. Again we did not find Hanson or any sign of him this day.

Thursday, March 13, 1975 6:00 A.M. - Day 5

We venture out into heavy rain today. It has warmed considerably with a high temperature of 50 and a low of 40. The search today will be in a one mile radius around Tricorner. Our team is assigned to search southeast in the Yellow Creek drainage toward Cataloochee as far as the Gunter Fork Trail. We divert down the old abandoned Dashoga Trail to look for signs. We find an old trail maintenance cabin we can utilize if the search shifts to this area, but no sign of Hanson there or anywhere.

Coming back to the Balsam Mountain Trail to Cataloochee we continue in that direction. We pass Thermo Knob named for an early surveyor who broke and left his thermometer there, finally reaching the Gunter Fork Trail. We drop down it for a ways but find no sign of Hanson. The rain gets even heavier as we continue our grid back toward Tricorner, where we are sure the camp custodian has dried our wet socks and boots in front of the fireplace. Darkness is closing in as I arrive at the shelter. I retrieve my boots and socks the custodian has dried for me and note he is sleeping soundly in his bunk. Tracker JR arrives to find his extra boots have been left too close and too long at the fire. The toes are curled straight up and it is impossible to straighten them. I ask him if he can bend his toes up for long periods of time so he can wear these boots. We have to restrain JR from grabbing the sleeping custodian by the throat. We did not find Hanson this day. He has been missing now for 4 full days.

Friday, March 14, 1975 6:30 A.M. - Day 6

Rainfall of 3.87 inches over the past three days and flooding is extreme. It will turn cold tonight with a low of 15. The search widens today with the arrival of two large Chinook helicopters landing three miles east of Tricorner at Deer Creek Gap. We search the Mt. Guyot area, the coldest place in the Smokies, in a grid pattern and also the Old Black Mountain area. The going is slow due to many blowdowns. We return to Tricorner at nightfall and again have to restrain tracker JR from throttling the camp custodian since just the sight of his desiccated, twisted boots sends him into a rage. Again no sign of Hanson was found today.

Saturday, March 15, 1975 6:00 A.M. - Day 7

We again experience cold temperatures and light snow as we are joined by 200 searchers arriving via helicopter. We grid search the Big Creek Drainage and Buck Fork Drainage which have been searched extensively on previous days. Keeping the grid in line is a major problem for us all day as we search farther and farther down these drainages and the slopes become steeper and rougher. We return again to Tricorner at dark without finding a trace of Hanson.

Sunday, March 16, 1975 7:00 A.M. - Day 8

A large number of searchers arrives again and we expand our grid farther and father down Buck Fork and other drainages in the area. Searchers on the Big Creek Drainage are suffering from expo-sure and exhaustion and have to be extricated via jeep which has to be driven above 3200 feet elevation. We do not find any clues, tracks, or any thing else on this day. Our search leader requests a pick up by the Chinook at Tricorner. We clear away from the shelter as the Chinook moves over the shelter area. A jungle penetrator is lowered and our fearless leader seats himself on it. High winds sud-denly appear dragging the leader on the ground first east then west as the Chinook tries to correct. I observe the leader has two pockets full of gravel when he is finally pulled aboard the Chinook. After a pow-wow and some minor first-aid, the leader returns with Monday's plan. We do not locate Hanson this day.

Monday, March 17, 1975 6:00 A.M. - Day 9

Rain, sleet, and snow fall on searchers today as we extend our grid farther down into Buck Fork. At approximately 2:00 P.M. searchers locate Hanson nearly two miles down Buck Fork Creek from the AT. He is sitting in an upright position leaning against a rock and tree. His parka is open, his gloves and a boot are in the snow nearby. Seven days have passed since Hanson disappeared. Doctors determine that he expired within 24 to 36 hours after he was last seen, when temperature was extremely cold and heavy snow was falling. We arrive at the site soon after he was found to assist in removing his body. Hanson's frozen remains are enclosed in a body bag which is placed in a litter and lifted by the Chinook helicopter.

Back at Tricorner shelter we gather our gear for the trek home. Most of the searchers are quiet and subdued. Considering the weather, time elapsed before Hanson was found and the circum-stances, the outcome of this search is not at all unexpected. At the same time, there is always disappointment and a feeling of loss when days of intense effort and hope result in failure to save a life. Lost in personal thoughts, we pack our stuff and get ready to go home. There are a few farewells.

When everyone is ready, we hike to Deer Creek Gap where we are picked up by one of the Chinooks. A hydraulic line blows over-head during liftoff, throwing occupants into a panic. The pilot says,

"This thing has two of everything, so do not be concerned." We try to relax. I note JR eyeing the camp custodian again and I move to his side to divert his thoughts during the short trip to the staging area at Cosby and to our parked cars.

It seems that youth is spent continuously pushing our envelope of experience. There is then not only boundless energy, but a feeling of invincibility as well. It is a time of risk-taking. Sometimes limits are learned during that process. At other times, good fortune smiles, rescuing the risk-taker from ultimate disaster. However, there are occasions when the intersection of hormone-driven euphoria with the reality of human physical limitations in adverse natural conditions, carries the individual beyond the limits, creating the nexus of tragedy. So it was with Mark Hanson. He over-reached without even suspecting his peril and thus crossed over the edge into the realm of mortal danger. The seductive irrationality of hypothermia blocked any hope he might have had for realization of his danger and, perhaps, of rescue in time. Even the good advice of his friend and companion failed to save him.

CHOPPER DOWN • BILL ACREE

A rented Cessna 421 airplane departs Fort Lauderdale, Florida at 4:00 P.M. on Tuesday, January 3, 1978, en-route to Chicago. It is piloted by Fred Philp, 27, and on board as passengers are his parents, Thomas and Elaine Philp, his brother Tom, aged 10, and his friend Marya Yates. The weather is clear and cool as the twin-engined plane which seats eight climbs quickly to cruising altitude. The trip is routine and the plane's occupants are returning to their home in Matteson, Illinois near Chicago after a winter holiday in south Florida. Over the next three hours, their flight plan carries them up the east coast of Florida, eastern Georgia, east of Atlanta and over the Great Smoky Mountains to Knoxville, Tennessee where they plan to land to refuel. Knoxville air traffic control has radio contact with the plane at 7:20 P.M.. The pilot reports flying at 8500 feet altitude in overcast skies southeast of Knoxville, requests and receives landing instructions. Only minutes later, the plane disappears from radar. Repeated attempts to contact the pilot afterward are unsuccessful. The incident is reported as a presumed crash and a Civil Air Patrol search is immediately organized to try to fix the location of the downed plane. At 11:30 P.M. search planes pick up the plane's Emergency Locator Transmitter (ELT) crash signal in the vicinity of Gregory Bald in the park. However the plane cannot be spotted in the darkness and there is no information about possible survivors. Radar tracking equipment showed the plane dropping off the radar scope within a mile of this location. It is too dark and the weather is too cloudy for either ground or further air search tonight. Search and rescue efforts are organized for dawn the next day to locate the crash and to extricate any survivors among the five known occupants of the plane.

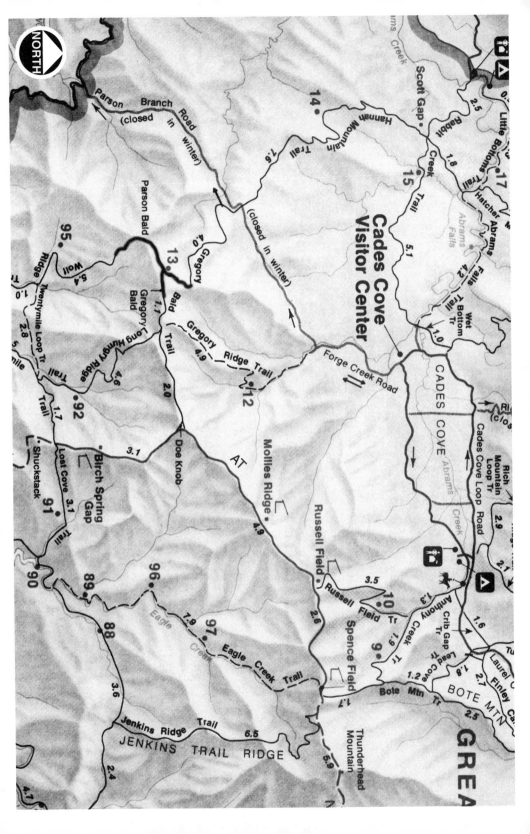

THE JOURNAL

Wednesday, January 4, 1978

*I am called just after midnight about the crash. Knowing the
area around the apparent location, I offer to hike in to see if there
are survivors who might need immediate assistance. Since it is dark
and the location is uncertain, the offer is judiciously declined and
I am assigned to search operations at Cades Cove where I am to
report at dawn. I arrive at the helispot at 5:00 A.M. where we gather
to meet a total of five helicopters called in for the search.*

*There are about 100 searchers here and we line up rescue vehi-
cles with headlights on to indicate the landing area for the choppers. Even before first light, two big olive drab UHI (Huey) helicopters from Ft. Campbell, Kentucky come in followed by another
Huey from the City of Knoxville, blue in color. Still another Huey is
from the Tennessee Civil Defense in Nashville. The last ship, a
smaller Bell Jet Ranger, is the medical rescue chopper from the
University of Tennessee hospital with Dr. Robert Lash and his
medical team. In addition, volunteer Civil Air Patrol planes are to
assist in pinpointing the crash site.*

*All helicopters arrive on site and shut down, waiting for day-
light at 7:47 A.M. when some will search and some ferry ground
searchers to the staging area which has been established at Gregory
Bald. The day breaks clear, sunny, and cold, with a temperature of
about 15 degrees. Weather on the mountain last night where the
plane went down reached a low of three. A very cold night if you
are injured and without shelter.*

*Nearing sunrise, all the choppers fire up their engines except the
blue Knoxville chopper which will not start. I decide that I do not
want to ride that one even if it does get started. At dawn some
searchers are dropped at the Gregory Bald area and begin making
their way toward the plane. Later when a CAP plane passes overhead it reports that the ELT transmitter on the downed plane is
being turned on and off, lending some hope that this indicates that
there are survivors at the site.*

We board the remaining helicopters and head out to Gregory

Bald area. The lead helicopter, one of the two Hueys from Fort Campbell spots the crash very quickly where it is spread over the mountainside just below Parson Bald on the Tennessee side of the Mountain. There are signs of fire at the crash site, but no outward signs of survivors. On board the Huey are the pilot, Captain John Dunnavant, Captain Terrance Woolever, a division medical officer and Sergeant Floyd Smith, the copter's crew chief. Also aboard are Colonel Ray Maynard of Knoxville, a logistics officer for the Tennessee Civil Air Patrol, Sergeant Chris Wyman, US Army paramedic from Fort Campbell and Sergeant Phillip Thurlow from McGee Tyson Air Base in Knoxville as well as Bill Acree a National Park Service Ranger and Dave Harbin, park technician, both from Great Smoky Mountains National Park.

Captain Dunnavant circles the site a couple of times and then banks to head toward the landing zone at Parson to drop the other personnel to search for and assist any survivors. It is only a few minutes into the flight and just 8:00 A.M. The chopper is climbing hard toward the summit and banking around toward the Bald for the drop when there is a loud report. The Huey suddenly loses power and veers to the right over the headwaters of the left fork of Moores Spring Branch just under Parson's Bald. The intercom crackles with the voice of Dunnavant shouting , "Jesus we've got an engine out."

The crew chief, in the right "bell-hole" seat with the park service personnel, shouts to the others, "we are going down, get ready." Everyone checks their seat belts and then settles in to watch the ground coming up fast as the pilot fights to find a clear spot to crash land. With almost no glide angle at all and no clear space below, the chopper quickly drops into the trees. The rotor blades catch hold of the heavy timber and rotate and flip the craft around and over. Strangely, the crippled engine does not stop running but continues to shriek at high speed after the crash. Inside, Bill Acree slowly regains consciousness to find that his right foot is pointed behind him and his whole lower leg is cocked at an odd angle. He checks to find that he has broken collar bones, probable broken vertebrae, broken ribs and other injuries. The helicopter is inverted and he is blocking what is now the only way to exit the broken aircraft, through a broken window which is now below him.

From the sounds of agony which he can barely hear above the engine noise, it is clear that some others on board have survived but it is hard to determine the situation exactly because of the attitude of the cabin and the scream of the engine. Somehow, driven by the danger of fire from the running engine and the need to clear the path for others to exit the craft, Acree forces himself to drop down through the splintered window of the cabin. He is able to slither on his back a few feet away from the twisted ship before getting hung up on some tree roots which prevent any further movement.

With the way clear, Harbin, who is also injured, but somewhat less seriously, is able to assist Wyman, who has a shattered leg, out of the window and away from the crash site before returning to assist Acree. Having salvaged the field radio, the three survivors request assistance to try to save the others, who appear to be either dead or unconscious, before the wreck catches fire. They do not know that Thurlow, also with numerous serious injuries has crawled through another shattered window on the other side of the craft and lies in a semi-conscious state in the undergrowth there. There is no sound or movement from the other occupants of the helicopter.

The pilot of the helicopter behind the Huey sees the crash, then turns immediately toward Parson, where he lands to discharge the searchers who now will be an emergency rescue force for those in the downed helicopter.

I am among those dropped off at Parsons Bald and go immediately to the crash scene below the Wolf Ridge Trail. While we are rapid-biting down to the Huey, other searchers arrive at the plane crash site and report by radio that there are no evident survivors at the scene. From the ridge above the chopper crash, we observe a helicopter hovering over the crushed Huey and one person rappelling down to the site. That soldier injured his back as he descended through the trees.

The injured soldier assists Harbin in tending to the survivors. He also determines that the others have not survived and cannot be *helped. We arrive some time later, and I note the jet engine is still running and making a horrendous noise so we move survivors to a location safe from fire upslope from the Huey. Doc Lash and the paramedics are tending to the victims while I take a chainsaw and cut a large area for the other chopper to winch out the survivors.*

Harbin's turn to fly out is last and he comes to me and says "Dwight I'm a little shy of riding another chopper out just now. Would you walk me out." I assure my long time friend I would but his injuries were too much for him and he finally elects to be lifted out. With all survivors gone to the hospital I note the jet engine is still running two hours after the crash. It finally runs out of fuel and stops.

I look at the Huey lying on its top and crushed flat in front. The rotor is in the top of a tree near the creek. Guards are posted at both crash sites as we prepare to leave the site for today.

Walking up to Parson Bald to be flown out, my mind wanders back to two other plane crashes within one mile of the present tragedy.

We returned to the crash area the next day, January 5th, for some final efforts on behalf of our comrades and of the strangers as well. This was a melancholy day as we hiked in to remove the fatalities from the crash sites. We used mules to carry the remains to a place we had cleared on a small ridge east of the crash and from there the bodies were airlifted out via a huge twin rotor Chinook helicopter. I was shocked at the utter destruction of the Cessna. It had first hit the ground, apparently under full throttle, on the old Appalachian Trail and sliced the forest for a quarter mile, ripping through trees and coming to rest in a little hollow. The occupants were strewn all around the crash site as the plane tore apart. It is tragic to see, for if the plane had been only 50 feet higher or a few hundred feet to the left it would have cleared the ridge and nine people might still be alive today. How the plane got from the 8500 foot altitude reported to the 3800 foot elevation where it crashed remains a mystery.

After removing the bodies we flew out of the area and I've

never been back. The wreckage of both crashes was removed at a later date. I returned to my normal duties and was glad to hear the four survivors would recover.

Bill Acree was in surgery for hours and in the Intensive Care Unit for several days. He suffered a broken back, a shattered right leg, two broken collar bones, broken shoulder blade, eight broken ribs, punctured lung, fractured right hip joint, dislocated jaw, broken teeth and multiple cuts and abrasions. He also got serum hepatitis from tainted blood transfusions, blood clots in his right lung and required bone grafts to mend his broken leg. It was more than a year before he was fit enough to return to his ranger duties.

My old friend, Dave Harbin, suffered from a broken rib, bruised lung, torn ligaments, dislocated shoulder, broken elbow and torn muscles. He recovered faster than Bill but is now no longer alive I am sad to say. He succumbed to a form of hepatitis a few years back, I believe in 1993. He was a fine lad and a joy to be around. I have been up the mountain and down the hollow with him and recall him as a man "with the bark on."

To this day, I still have occasional regret at not having pressed harder for permission to search for the Cessna crash site the night it went down. If I had tried it and had been successful, we would have known of the demise of the passengers and might not have mounted the massive helicopter rescue operation the next day, thus saving four lives.

GONE FISHIN' • ALBERT BRIAN HUNT

The sun was just coming up over the horizon behind his right shoulder as he swung his nearly new 1980 Mazda 626 up onto the blacktop of Highway 27, headed north. It was one of those silver-mist Florida mornings, April 29, 1981. Twenty year old Albert Brian Hunt was on his way from the family ranch where he lived with his mother, Pattie, just outside Arcadia, for a week or so of camping and fishing. The light blue two-door sedan, was smooth and responsive, accelerating easily to highway speed as he settled in for the long drive north. The country here, just at the edge of the everglades, is flat and featureless. Just the kind of place for a cattle ranch like the one run by Hunt's family, but a sharp contrast to the country for which he was headed.

By early on May 1, moving at a vacation pace, he had gotten as far as Macon, Georgia where he charged a tank of gas at a Chevron station on Interstate 75. From there, he left the interstate highway system just northeast of Atlanta and followed secondary roads into the Nantahala National Forest where he spent the rest of that day and night. The following day, May 2, he checked into campsite D-41 at Smokemont, sharing the location with two others.

His companions, whom he met by chance in Cherokee, the eastern gateway to the Great Smoky Mountains National Park , were Robert Elton Taylor and Freddie Ray Staton. They were cousins, about ten years older than Brian, and currently hailed from near Lawrenceville, Georgia. They were drifters and itinerant workers with a passion for beer, pot, and easy living. When they all went together to get gas and supplies in Cherokee, Taylor and Staton noticed that Hunt had several hundred dollars cash in his wallet. The three spent the day together and talked about their common love for fishing. The other two told Hunt they knew a special place in the park where the fishing was outstanding and they would take him there the next day. Early on May 3, they left their camp at Smokemont and drove over Newfound Gap to the Tennessee side of the mountains to the place just below the gap where Walker Camp Prong crosses the road. Here they parked the

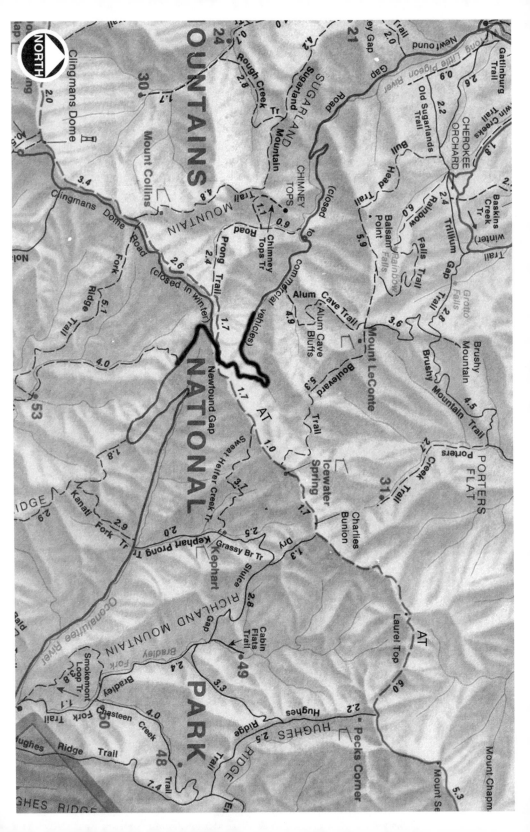

car near the stone bridge, took their fishing gear and started hiking up the creek.

At first, the going was pretty easy where the vegetation had been trimmed back by park maintenance and trampled by heavy use. However, they were soon surrounded by dense undergrowth of laurel and rhododendron which overhang the channel of the stream. For almost a mile they fought through the thick growth and steep grade to a fork in the creek and then took the left branch looking for the ultimate fishing hole at a place just above a little waterfall. Staton and Taylor had carefully planned the next move the night before and, at an opportune moment, they jumped Hunt. Although surprised by the move, Hunt, a stocky 5 foot 10 inch 170 pound guy in good physical condition, at first held his own against the two. However, he made the fatal mistake of letting Staton get behind him. With a large rock for a weapon, Staton hit him on the back of the head and he went down bleeding profusely. The two went through his pockets for his wallet and keys and then left him dead or dying at the edge of Walker Camp Prong. They hiked out to the road where they got into Hunt's Mazda and drove away.

Later the same day, they were back in Lawrenceville where they visited an acquaintance and tried to sell the car for $400 without the title. This effort failed, but they did dispose of some of Hunt's camping gear and books. Other attempts to dispose of the car were also unsuccessful so they drove the car to a remote area in Gwinnett County near Dacula and burned it.

Not having heard from her son in several days, Pattie C. Hunt contacted the police in Arcadia at almost the same time the partially burned car was discovered. A search bulletin was issued for Hunt and authorities in Georgia, where the record of his gas purchase in Macon was soon found, began a massive search for him. Several weeks went by while Gwinnett County police, the FBI, and other authorities tried to first find Hunt and then find out what happened to him. Although Taylor and Staton had attempted to remove identifying marks on the car, investigators soon were able to trace it to Hunt. Inquiries in and around Lawrenceville led them to Taylor who had been very open in his attempts to sell the car.

By June 4, 1981 they were ready to arrest Taylor on charges of auto theft related to his possession of Hunt's car.

Faced with these charges, Taylor is said to have been offered an interesting proposition by the police. If he helped them solve the mystery of Hunt's disappearance, they said that they would not prosecute him for auto theft. Thinking that seemed like a good deal, Taylor made a voluntary statement confessing to the assault and murder. He also implicated Staton who had already fled the state, and provided much of the detail of the crime to authorities. Somehow, he was surprised when the police later arrested him for murder.

On July 10, 1981, Gwinnett County police and FBI agents from Knoxville together with Great Smoky Mountains National Park personnel made a field excursion up Walker Camp Prong to the crime scene led by Robert Elton Taylor. The scene was bizarre with park rangers in mufti, blue-uniformed police, plainclothes detectives in casual street garb and FBI agents in wing tip shoes and black or blue suits all crashing through the underbrush and splashing through the creek for more than a mile to the scene of the crime led by a perpetrator who was not only helpful and cooperative, but actually garrulous.

THE JOURNAL

Thursday, July 9, 1981 4:00 P.M.

Today, I am assigned to participate in a search for the remains of a young man who is said to have been killed somewhere just below Newfound Gap, maybe on Walker Camp Prong. The FBI and law enforcement personnel from Georgia will be coming in tomorrow morning for the search.

Friday, July 10, 1981 7:00 A.M.

I join another tracker and we meet FBI agents at Walker Camp Prong Creek in the park. One man who seems to be a suspect leads us one mile up into this wilderness area where there are no trails and no one goes. This is the same area where Geoff Hague died

*eleven years ago. Agents in their blue suits and wing tip shoes have
a hard time climbing up the many waterfalls on this creek and one
damages his pipe in a fall. After climbing the creek for about a mile,
the suspect stops and points to the place where Hunt was last seen.
I gather from the conversation that the suspect and a companion
befriended Hunt and talked him into going to this remote area,
and then did him in.*

*We get up there and climb this little waterfall. On the other side
of it there is this huge rock and next to it a small rock that they had
evidently used as the weapon. "Right here is the place; this is where
we killed 'im," says the suspect. But he is not here. He is gone! One
of the FBI boys says, "You are lying, where's the body?"*

*"This is where we done him, here's the rock," says the guy. At
first, there is not much to be seen here, but I know what has probably happened. While the others are arguing about the missing
body, I ignore the strange looks of some of the others and look
around really good on my hands and knees with my nose almost
on the ground. Soon I see a yellow-white gleam under a log across
the creek which turns out to be a lower jaw bone. Then, soon
enough, I unearth other human bones and clothing. Although only
a few weeks have passed since the killing, all of the bones are
stripped clean and shining. There is no sign of flesh or blood. As we
find each item, it is tagged and saved in a plastic bag.*

*Our search and eventual finding of numerous bones leads us
uphill a quarter mile from the initial crime scene. The area is thick
with rhododendron where you can only see a few feet in front of
you. In nature nothing is wasted and that is the case here. When
an animal or person dies in the wilderness, after a period of one to
two weeks passes and, if the body is not disturbed, it then becomes
a feeding area for bears, bobcats, possums, birds of all description
and, finally, insects. Bears are at the top of the food chain due to
their size and disposition. From the signs in the area, I observe that
five bears of differing size had been here, two big ones and three
smaller, maybe yearlings. The larger bears chased the smaller ones
up trees where they left five different sets of front claw marks in the
bark. The big ones evidently fought over the body and in doing so,*

separated it into two parts. The upper part wound up below at the death scene and the other up here in the rhododendron thicket. Up here we find the victim's pants and belt buckle, several vertebrae and leg bones.

We should not be shocked by the rules that exist in nature and should be well prepared to live by these rules when we visit the wilderness. I am glad that today's date is not May 10th instead of July 10th, for if it were so we would be in a confrontation with five bears. Once they establish possession of food they give it up reluctantly.

After we tag and bag all remains and evidence the bags are put in a backpack which I carry. It is large but not heavy. We then hike out back to the highway where I walk by several news reporters with their video cameras waiting to shoot a gruesome story. They ask the agent, "When will the body be brought out?" He replies, "It just passed you." I am already in the truck and on my way, so there will be no tape for the news today.

We located Hunt's remains at 1:00 P.M. and brought him out of the wilderness at 6:00 P.M. I hope this will grant some closure to the family who have suffered for over two months not knowing what happened. I also believe we can use what we have learned from this search to benefit other searches.

On July 21, the FBI laboratory positively identified the jawbone from dental records and other forensic information as belonging to Albert Brian Hunt. A nationwide search for Freddie Ray Staton located him in Bureau County, Illinois where he was arrested and held on a charge of second degree murder on August 9, 1981. He waived extradition and was returned to Knoxville for trial. In his statement, he admits to the crime but claims that Hunt was struck with rocks by both men and that their intent was only to render him unconscious and to rob him, not to commit murder. Based on this information, on August 27, FBI agents returned to the crime scene with park personnel for additional search, finding a large part of the skull intact and several fragments as well. These were examined by Smithsonian Institution experts who concluded that there is evidence of fracture from a blunt weapon such as a rock.

On November 10, 1981, a complaint was filed in US District Court in Knoxville, Tennessee against both Robert Elton Taylor and Freddie Ray Staton charging the two with murder, robbery, and interstate transportation of a stolen motor vehicle. The complaint and evidence was heard by the Grand Jury which returned an indictment on November 20. Both defendants had attorneys appointed by the court and were held for trial. At the request of his attorney, Staton was sent for psychiatric evaluation and was determined to be legally sane and fit to stand trial. Attorneys for both reviewed the evidence against them and entered into plea bargaining with the federal prosecutors.

After extensive negotiations, both men agreed to plead guilty to second degree murder and the other charges in exchange for a 35 year sentence. Taylor was sentenced on February 5, 1982 and Staton on April 23, 1982. Both have since petitioned the court for reduction in their sentences based on grounds that the victim was merely rendered unconscious by their attack and that the bears, because they subsequently killed and later dismembered and ate the victim, should be held responsible instead. To date, their appeals have been unsuccessful and both are still in federal prison.

A PERSONAL QUEST • John Hernholm

John Hernholm left his home in Memphis on an odyssey. He had been out of work for several months and was doing a lot of soul searching as a result. Borrowing some equipment from girl friend, Sharon Stroud, and some from another friend, he left Memphis on February 20, with a promise to return by March 9 with a new outlook on life. He called her on the 23rd from Missouri and said he planned next to be in Arkansas. He may have then visited his parents in Baton Rouge before driving across Tennessee to the Smokies and the next part of his personal quest.

When he arrived at Sugarlands Visitor Center, he was planning a six day backpacking trip on the Appalachian Trail (AT). His permit was to stay at Icewater Springs Shelter on February 25, Pecks Corner on the 26th, Tricorner on the 27th, Pecks again on the 28th, and back to Icewater on March 1. He was hiking alone but had adequate equipment for a normal winter hike. He had hiked several times before and camped in this part of the Smokies on at least one previous occasion. He is a big guy, over six feet tall and about 225 pounds. Although not really athletic, he is in reasonably good shape for his 30 years. Three years before this trip he was partially paralyzed on his left side from an unknown cause, perhaps a stroke. Although it left him a little less coordinated, he is slowly recovering physically. He is also depressed over his inability to find another job to replace the one he has lost a few months before.

Leaving his car in Gatlinburg and carrying all of his gear for the trip, he hikes up Route 441 to Newfound Gap and then the three miles east along the Appalachian Trail to Icewater Springs and the shelter. Here, by chance, he is joined by another winter hiking group from North Carolina, three friends, Rusty Poole, Chuck North and Chris Brocknail. The four young men exchange cordial conversation for awhile and then turn in for the night.

At dawn about 7 A.M. the next day, John and the Poole group

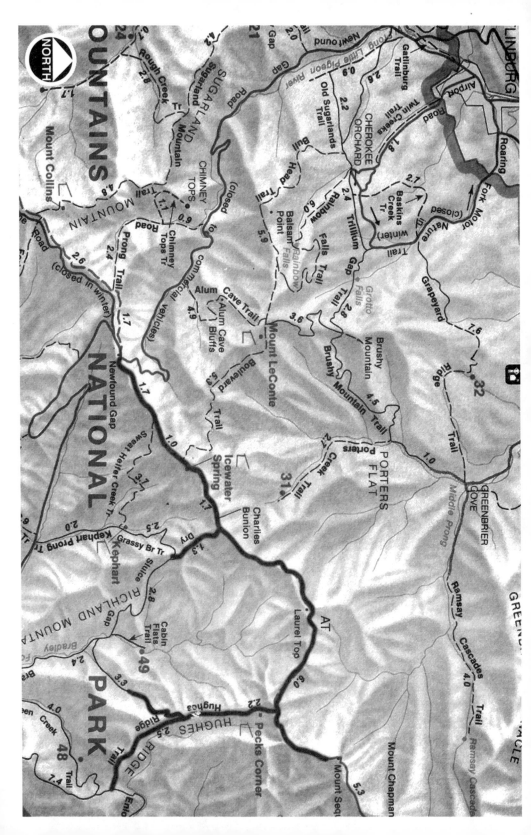

all wake, prepare breakfast together, and get ready to break camp at the shelter. The temperature is in the low twenties and it has begun to snow. The Poole party is headed for Tricorner Shelter, about 12 miles from Icewater, for their next night. Since Pecks Corner, John's next scheduled bivouac, is about 8 miles along the same route, they all start off east together. Less than two miles from Icewater, at Charlies Bunion, they stop to rest briefly and then continue to march eastward in increasingly heavier snowfall and rising wind.

Fighting the wind and the snow, the rest of the group is moving faster than John can go under these conditions and he begins to lag behind so he tells them to go ahead and he will catch up. By the time they reach Porters Gap, they have lost sight of him but expect to see him again later. By the time they reach the trail junction at Pecks Corner the now howling storm forces the Poole group to scratch their plan to go all the way to the shelter at Tricorner Knob. They decide to wait out the storm at Pecks. To be sure that John doesn't change his plans thinking they will be at Tricorner and try to push on himself to the shelter there, they leave a note for him on the trail sign at the trail junction of the AT and the Hughes Ridge Trail indicating they will see him at Pecks. They arrive there about 1:30 P.M. and wait for John to catch up.

At about 3:30, another party, led by David Whitmore, arrives at the shelter and, after a brief discussion about John, some of the combined group walk back up the trail to the junction with the AT to look for him and to gather firewood. There is no sign of John Hernholm. Not knowing whether he might have turned back or taken shelter somewhere along the way, and with approaching darkness, they return to the shelter and wait for him to appear. They prepare dinner about sunset, talk for a while and then retire for the night, hoping for the best for Hernholm.

The next morning, Saturday the 27th, the Pooles awake before dawn to a heavy snow. It has accumulated to a depth of eighteen inches now with drifts as deep as three or four feet. They are all now really concerned for John who still has not showed up. As they retrace their route to Icewater they note their own footprints from the day before are not visible nor are any more recent footprints. They are breaking new snow as they travel. They get to the trail junction without any sign of Hernholm and begin to trek west toward Icewater to see if they can find him. About a half mile west of the junction on the Appalachian Trail they are stunned to find, almost completely buried beneath deep snow, backpacking equipment they recognize as belonging to John. His orange sleeping bag is unrolled and lying across the AT. His closed-cell foam mattress pad is unrolled but not under the bag. There is also a canteen partially full of ice.

Shocked and deeply worried by what they have found, the Poole/Whitmore group pick up John's equipment and carry it with them. They continue to hike, now very urgently, toward Icewater making slow progress through the deep snow. When they reach Charlies Bunion without finding John they decide to leave his equipment in a sheltered place and rapid-hike out to report Hernholm missing. The going is very difficult due to the depth of the snow and the continuing storm. They finally reach Icewater Shelter about 3:30 P.M. where they encounter two other parties coming from other directions. None had seen any sign of John. Exhausted from their hike through the snow drifts, they spend the night here and leave the next morning at 7 A.M., reaching Newfound Gap where they report Hernholm missing to the ranger on duty.

THE JOURNAL

Sunday, February 28, 1982 8:30 A.M. - Day 2

Hasty-Search parties are immediately dispatched to re-cover the trails from all directions into Pecks. One other searcher and I are called to work as a team and we respond with overnight gear for assignment to hike in from Newfound Gap. We have 10.4 miles to hike to Pecks in the storm. We look for tracks and any other indication that John may have left the trail. The snow has covered and obscured any tracks in the area even those of the Poole/Whitmore group. At False Gap, we visit the old shelter site at the spring on the

Tennessee side. The going is physically demanding and slow; other search teams are having the same difficulty. It is almost 5:00 P.M. before we crest Laurel Top Mountain. I give a close look a the Tennessee side of Laurel Top to see if a person has gone in the direction of an old plane crash that occurred a short distance off the AT. The plane was a D-17 Staggerwing Beech listed missing 8/12/44 with 3 persons aboard. The wreckage was found 9/19/47 by hikers. The three occupants of the plane were never found.

Having determined that no one had ventured toward the old crash, we continue on. At about 7:30, we locate the site where Hernholm's equipment was found and find his black and red ski glove which had been overlooked by the others. We know that aban-donment of his equipment is one of the signs of advanced hypother-mia. We continue on to the junction of the Hughes Ridge Trail where we search with flashlights for approximately one mile further east toward Tricorner on the assumption that Hernbolm went that way. There is no way to pick up any tracks in the deep fresh snow. I do try several times, clearing more than 2 feet of snow away to observe what lies beneath. I find a jumble of melted, then frozen then thawed then re-frozen, tracks of hikers. I have no idea what John's prints look like either. So we are basically looking for sign of a person breaking limbs off near the trail or leaving items or equip-ment behind.

There is no further sign of John toward Tricorner so we return to Pecks Shelter and search that area. Finding nothing there, we continue to the old maintenance cabin farther down the trail. By this time it is 10:00 P.M. and we are exhausted from struggling in the snow for over 12 hours. I build a fire in the old wood stove and light the gas lamp. I boil some hot soup and hot citrus drink and relax in the upper bunk reading a Louis Lamour western. My com-panion makes a sumptuous meal of a homemade concoction of hamburger helper. It looks and smells very delicious but he fails to offer me any. Within an hour I am actually grateful for the slight as he tells me he is deathly sick from the slop. Thereafter, every 30 minutes or so, I am obliged to assist him outside to relieve his dis-tress. Neither he nor I get any sleep the whole night.

Monday, March 1, 1982 7:00 A.M. - Day 3

We continue searching the immediate area and down Hughes Ridge Trail. Over the radio, we hear other searchers have found more of John's equipment two miles down Hughes Ridge Trail. Two search parties, having joined forces, and coming up the Hughes Ridge Trail, locate a faded blue backpack under a spruce tree about ten feet off the right side of the trail in the snow. Careful removal of the snow reveals some heartbreaking information. The main compartment of the pack is unzipped and almost filled with snow. Next to the pack, nearly buried, is a Coleman stove and close by, a fuel bottle which is empty and missing its cap. A couple of packets of freeze-dried food are also buried in the snow. Next to the stove it appears that an attempt was made to build a fire. There are small twigs laid into a pyramid. Alongside there are several stick matches scattered about unburned and a roll of toilet paper. There are also numerous personal and toilet articles strewn about. In an unopened compartment of the pack are more food packets and articles of clothing. The pack and everything else is wet. Hearing the description over the radio, I can think only of Jack London's To Build a Fire.

In the next few minutes many thoughts race through my head. If this is indeed John Hernbolm's gear, and it seems that it could be no one else's, how did he get down past the shelter at Pecks and the maintenance cabin just behind us, both of which are within 30 feet of the trail, without noticing either one? There were two parties of hikers in the shelter, a fire, lights, and voices, all of which should have been visible and audible from the trail even in the dark and in a storm. If John could see well enough to have found the trail junction of the Hughes Ridge Trail and well enough to have taken it at all, why would he have not seen the sign telling the short distance to the shelter and found it by distance alone? Even if he missed the shelter, how did he also miss the maintenance cabin which is even closer to the trail? Although it was locked, it would not have been difficult to break in and build a fire or to get warm in the straw which is stored in the little barn there. My mind is gnawing on these puzzlers as we continue the search toward the location of the other searchers and that of the latest find. The radio crackles with the latest stunner. Just as we are about to meet the other party a few hundred yards away from us now, we are told that they have found John. It is exactly 12:16 P.M.

Less than a half mile past the maintenance cabin down Hughes Ridge Trail the other search teams have found John Herndon lying across the middle of the trail. Partly covered with an orange poncho, and under a thick cover of snow, he can hardly be seen. He is lying on his back, invisible under the snow blanket except for the colors of his red and blue vest, blue jeans, and orange poncho which peek through. When the snow is brushed away, we can see that his right arm is bent at the elbow with the forearm and hand resting on his chest. His left arm is straight beside his body, his feet are crossed at the ankles, at rest. Both boot laces are untied half way down the boot. Under his vest he is wearing a patterned, dark, flannel shirt. There is a black and red ski glove on his left hand and nothing on the right. He is wearing gray woolen socks and brown suede hiking boots. A tri-fold wallet with his drivers license and other ID and two uncashed unemployment checks for 91 dollars each are in his right rear pocket along with a pair of pliers and a black leather knife case. The left rear pocket holds a blue ban-danna. About five feet north of the body, toward Pecks shelter, there is a blue ski mask almost buried in the snow. No tracks are visible other than ours anywhere near the body indicating that he has been here for many hours.

While the others finish recording the information found here, I walk on to the intersection of the Hughes Ridge with the Bradley Fork Trail. There in the shelter beneath a tree where the direction sign is clearly visible I find tracks beneath the new snow which approach the sign from the north and turn to go back in that same direction. In my mind's eye I can see John in the middle of the night, half frozen and near dead from fatigue, standing by this sign, realizing that he had missed the shelter and that he would have to climb the 1.3 miles back to it if he was going to survive. He must have tried hard to do it but missed it by less than a third of the distance.

I rejoin the others as we load a horse to transport the remains. It's a long hike out, approximately 12 miles, to the trailhead at Smokemont. The consensus of opinion is that John succumbed to the weather on 2/26/82 sometime during the night. We talk about what more could have been done, but it is clear to me that the response was both rapid and appropriate to the situation once we knew he was missing. Helicopters or any air search was impossible in the storm and vehicles were not usable in the deep snow. In the final analysis, it looks like he was dead for hours before he was even known to be missing. No search effort could have helped him then.

Friday, March 5, 1982

I am back at work when I hear that Ranger Gary Stewart was sent out to search for any other information about Hernbolm's loss. The snow has been melting the past few days and if there is anything else to be learned from this event, now is the time to find the clues.

As he walked down the Hughes Ridge Trail from the junction with the AT, Gary found a trail of heartbreak and delirium. About 200 yards from the junction there was an orange Techsport tent with stakes and poles in a tan carrying sack, still partially buried in the snow. We walked over this in the dark and didn't see it because it was deep in the snow.

About 100 yards further down and right on the trail was John's Explorer survival stainless steel knife which belonged in the sheath found in his right rear pocket. A little further along was his Kodak camera in a black leatherette case and near it, a Brewster briar pipe, well smoked.

From all of the clues found, the last few hours of John Hernbolm's life must have been very difficult. As the others got farther ahead of him, John probably found himself biting the knife-edge Ridge along the AT where the trail follows the crest of the ridge. Here, the biting cold wind and snow must have cut through him like a razor. After climbing Laurel Top Mountain, he begins to really feel the effects of hypothermia. He attempts to drink from his canteen but the water is frozen solid. He continues on, not realiz-

ing he has dropped his water bottle in the snow. A little farther on he removes his sleeping pad and drops it too. A few yards later, he discards his goose down sleeping bag as well. Stumbling on, he drops his right mitten in the snow.

John arrives at the Pecks Shelter turnoff and is able to read the sign indicating Pecks is down the trail to his right but darkness has overtaken him and it is around 9:00 P.M.. He hikes past the Pecks Shelter which is off the trail a little way on his left . The Poole party and the Whitmore group are there in the shelter talking and cooking but John does not see or hear them. John continues down the Hughes Ridge Trail. A quarter mile further he passes the maintenance cabin that is only 20 feet off the trail to his left. He continues on down the trail another mile to a trail junction and a sign. Here he brushes away the snow and ice and feels the letters routed in it saying Pecks shelter is 1.3 miles back up the trail he has just come down.

John starts back up the trail and about half way he stops and sets his backpack against a tree and tries to build a fire. Failing, he leaves his equipment and continues up the trail. Less than a quarter mile from reaching the maintenance cabin John lies down in the middle of the trail pulls an orange parka partly over him and loses his struggle with nature. The snowstorm is unrelenting and continues to heap layer after layer of snow over him and it covers all of the footprints he has left. John has come 12 miles in very bad weather through nearly two feet of snow and four foot drifts in places. The effort of pulling his shoes out of each step in the snow has taken its toll.

We are all, at times, balanced on a knife edge between life and death. Given the weight of his depression, his physical handicap, his pride, the weather and terrain, John Hernholm was no longer able to counterbalance them with physical strength, endurance, and will to live. He over-reached and, as the balance tipped against him, he slipped over the edge.

Stand By Me • Rick Callahan

On Monday, May 6, 1985 two young men from Indianapolis, Indiana arrive in the Great Smoky Mountains for a week-long backpacking and camping trip in the park. They check in at the Cades Cove ranger station and receive permits. Their plan is to spend the first night at Spence Field shelter and the night of the 7th at Derrick Knob about six miles east of Spence on the Appalachian Trail. Continuing east the next day they will stop at Silers Bald, another six miles or so. On the 9th they will turn off the AT on Welch Ridge Trail into North Carolina, hiking down Hazel Creek to Camp Site #84 where Hazel Creek Trail and the Lakeshore Trail join. Although this leg of their hike is a long one at a little over 12 miles, it is almost all down hill. From the camp site, they will follow the Lakeshore Trail west to the Jenkins Ridge Trail, turning north again and climbing back up to the AT to return to Spence Field Shelter for their last night. This leg is about eight and a half miles and mostly uphill. The trip is well-planned in that the first three days are pretty easy hiking, the fourth a bit longer but downhill, and the only hard part comes after they have had a chance to acclimate to the rigors of the trail.

They park their AMC Pacer and leave Cades Cove about 1:00 P.M. on the 6th, hiking up the Anthony Creek Trail to Spence, arriving there in plenty of time to set up camp, cook dinner, and get a good night's sleep. The routine and rigor of a backpacking hike in the mountains takes a little time to adjust to and they don't sleep quite as well as usual. However, both men are in their early twenties and in good physical condition. Their hiking experience is limited, but they have been out like this before and they are confident of their ability to make the trip. Walt Johnson, 23, from Greenwood, Indiana, is the more outgoing of the two, a little huskier and stronger, and often takes the lead in their activities together. His companion is Rick Callahan, 21, who lives in Indianapolis. Rick is tall, about 5 feet 10 inches and slender, about

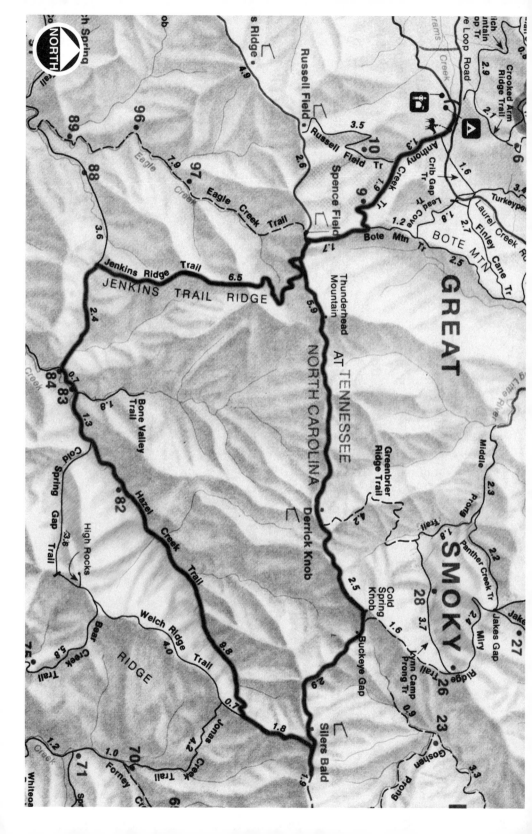

170 pounds. Both are students in their home state and eager to cast off the academic burdens at this the end of their school year. They are well-equipped for a late spring hike and they set off from Spence with some vigor despite their minor aches from the first day's hike.

Early on the 7th, they break camp at Spence and as the day warms, begin walking east on the AT toward their next night's shelter at Derrick Knob. They pass the trail fork for Jenkins Ridge and are climbing toward Rocky Top in the hazy sunlight when they stop for a short rest. They look at the map which Walt is carrying and, realizing that the trail follows the narrow ridge of the crestline of the mountains, they become discouraged at the up and down hill hike they can expect for the next two days, and decide to alter their plan. This is a deviation from their backcountry permit which requires them to follow the designated trails, using the shelters permitted. Nevertheless, they decide to take a shortcut from Spence to Hazel Creek and cut off the three days which would have taken them to Camp Site #84 via Derrick, Silers and the Hazel Creek Trail. Looking at the map and using his compass, Walt sights a line in a southeasterly direction toward the location he thinks is that of Campsite #84 in the Bone Valley-Hazel Creek drainage. Trying to keep his line in mind, they leave the AT heading east-southeast in open wooded country near the well-traveled trail.

Rick

Rick follows as Walt drops off into the head of Bone Valley. The terrain quickly turns ugly with a near vertical bluff and a very steep slope under thick brush and trees. Rick finds it nearly impossible to carry his backpack in the underbrush. Walt seems a bit frustrated with Rick's caution and slow pace. Rick continues to follow Walt as best he can. They are on the west fork of Roaring Fork Branch and soon see another stream has joined the creek they are following. Farther down they see yet another stream joining theirs from the left (Mineral Gap Branch). A short distance farther they pass still another stream this time joining from the right (Rock Camp Branch), where they stop to fill their canteens. Their route gets a little flatter and they are now on what seems to be an old road but is really an old logging railroad bed. It parallels the stream, sometimes right next to it and sometimes a little distance away. However, it is choked with underbrush and getting through it is no less difficult than bushwhacking right next to the creek except that it is drier. Rick follows Walt a half mile farther but Walt is steadily pulling ahead until Rick can no longer see or hear him.

Rick finds himself alone on Roaring Creek. His calls to Walt are unanswered, perhaps unheard against the roar of the stream. Rick comes to another creek entering from his left . He does not know that its name is Desolation. The sky is becoming fully overcast. With the changing light and unfamiliar surroundings Rick becomes confused and starts to wander about. He is alone. He is not familiar with the area. He does not know what has happened to Walt, who has the good map and compass. Could Walt be back upstream? Could he have left the rough trail to relieve himself while he, Rick, walked right by him, not knowing it?

Deciding that he must somehow have missed him, Rick turns and heads back up the creek he has just come down. Only this time, instead of following the old railroad bed, he follows the stream itself. Rick is not a natural leader and does not feel comfortable going into unknown areas. Reaching the flat area he passed previously, he decides to stop and wait for Walt to find him, feeling confident that Walt would not leave him. By this time, darkness is approaching and it is starting to rain. Rick makes a hasty camp as the rain comes down more heavily. He tries to build a fire which is too small and rain will not let burn. Wet and miserable, Rick settles down to spend the night. Climbing into his sleeping bag he finds his poncho has been shredded by his hike through the trees and brush and is allowing the rain to soak his bag. It is a dark and fitful night trying to sleep in the wet bag and he feels very much alone.

Walt

A little impatient with Rick for lagging behind and feeling energized by the downhill walk, Walt continues to move along down

Roaring Fork Creek on the old logging railroad bed. Almost without noticing it he pulls far enough ahead of Rick so that when he stops for a short breather, he can no longer either see or hear Rick behind him. Thinking his companion will soon catch up, Walt continues to move downstream for awhile, although slowing his pace a bit to assure an eventual reunion. When that doesn't happen, Walt again stops to wait for Rick, this time for a longer period. When he still does not appear, Walt turns around and retraces his steps all the way back to Rock Camp Branch, the place where they filled their canteens.

No Rick. Walt again turns around and hikes back to the place where he originally started to retrace his steps to search for Rick. Still no sign of him. Feeling very anxious by now, Walt drops his pack and starts to run up and down the stream calling for Rick. Frightened that he may have fallen into the water or injured himself on the rocks which jumble the bed of the stream, Walt plunges into the water recklessly looking for some sign of his friend. To make things worse, rain begins, and whatever parts of his clothing not thoroughly soaked already from sweat and his sorties through the stream, now become wet with cold rain.

In the mountains, night falls rapidly and there are few precious moments between the sun dipping behind the peaks and pitch black. Walt sees that it is getting dark now, and he decides to retrieve his pack and try to make a camp for the night. In his frantic search for Rick, he forgets where he left his pack. Now he's in the rain, wet, cold ,and without matches, sleeping bag, tent , or food. It is now pitch dark, and there is nothing to do but wait for light again. Fortunately it is nearly summer and warm. He removes some of his clothing to try to dry it, finds a relatively dry spot under a tree and settles in for the night.

Rick • May 8

Rick awakens in the morning and Walt has still not returned. His sleeping bag and sweater are soaked. He leaves his sweater with its stuff sack, a Frisbee, his right tennis shoe, and several food items. Rick decides to return to Cades Cove the way Walt had brought him, down the creek. He leaves a note to Walt which says, "Walt I am lost and will go back upstream to Knoxville." Rick attaches the note to a tree and covers it with what was left of his poncho. He tears out the pocket of the orange poncho and fills it with pieces he has torn from the shredded cape. He packs up his wet equipment including the stuff sack containing tent poles to Walt's tent that Walt still has.

Rick leaves his camp and starts flagging his route with the poncho pieces. He walks upstream three quarters of a mile to the left fork of Roaring Creek which he and Walt came down the previous day. After traveling a quarter mile up this left fork, the difference between hiking downhill through this dense underbrush and doing the same uphill becomes painfully clear to Rick. He decides to lighten his load and discards Walt's stuff sack with the tent poles and his own sleeping bag which is soaking wet and very heavy . He then turns back downstream after leaving the items on a log in the middle of the stream. Twenty feet below the log Rick discards the red tie string for the tent sack. Panic is draining his energy.

After reaching the creek junction, Rick turns up the right fork hoping it will be easier going. He continues to flag his route. He goes for a third of a mile and stops to eat an apple. Continuing on, Rick does not notice that he has left his carrots, fruit rollups, and several other food items lying where he sat. A quarter mile farther Rick loses the poncho pocket full of pieces he has been using to mark his trail. He does not know it's missing. Rick comes to a three way split in the creek and decides to take the left fork. The drainage becomes extremely steep. After traveling some distance, Rick loses or discards a large container of drink mix, a six ounce can of tomato paste, Italian food in a zip lock bag, and several other food items.

Farther up the creek, the brush tears away a portion of his official NPS backcountry map but Rick does not notice as he is having a hard time climbing the steep brushy slope. He turns right into the thick brush and sidehills to the right hoping to get clear of it. At one point, he discards his pack which has gotten hung up in laurel. He puts his remaining possessions in a laundry bag which is easier for him to drag through the brush. He inventories what he has left, a blanket, a first aid kit, some food items, some clothing, and fire starting material. The going is better now but darkness has again caught him and he camps in the brush, on a ridge.

Walt • May 8

Cold and still wet at the first glimmer of dawn, Walt starts moving around to search for his pack and for Rick. He finds the pack after a bit, hurriedly eats some dried fruit and nuts from it and begins again to search for Rick. He wanders about without any plan or system between the place he last saw his friend and the lowest point he reached in his own trek. By early afternoon he has had no success and decides to continue downstream in hopes that Rick may have somehow gotten by him and is somewhere farther down the creek. Still following the old logging railroad bed, he eventually comes to a place where there is an old cabin (Hall's Cabin) in a clearing with a faint track of a jeep trail leading to it from downstream. Next to the cabin is the foundation of another building. Inside the front door there is a room to the left and a set of stairs leading to a loft or attic space above. Mindful of his experience the night before when he was caught unprepared by nightfall, and with his clothing still wet, he finds wood, builds a fire to dry his clothing, and makes camp inside the cabin. Night finds him fed and relatively warm and dry inside Hall's Cabin on the Hazel Creek Trail.

Rick • May 9

Rick continues this morning traveling along a very rugged ridge still moving to his right (east). He has slowed down considerably and does not make good time. He stops at 2:00 P.M. and gathers as much firewood as he can find to ward off the night chill and the strange sounds that frighten him at night. He feels the woods are full of unknown creatures. He huddles close to the fire he has built, continually feeding it throughout the night. But by 2:00 A.M. all the wood is gone and the fire dies. He hears what must be a snake come near him and hears grunts and whooshing sounds close to his camp. Up in the trees close by there is an ear piercing shriek that echoes in the hollows all night.

Walt • May 9

Walt wakes more rested than he was after the previous night, but still very anxious about Rick. He fears that he will not find him alone and that help is needed. After a hurried breakfast, he looks at his trail map and decides on a course of action. He packs his gear, shoulders the pack, and heads out southwest on the Hazel Creek Trail, takes a right at the junction with the Lakeshore Trail and another right at the Jenkins Ridge Trail. It is a challenging hike, mostly uphill, as he climbs from an elevation of just over 2000 feet to over 5000 feet where he rejoins the AT. However, he is energized by the need to find Rick and by his own decision to seek help. He barely pauses when he crests the ridge top of the mountains, having already come over nine miles on a steep grade, and plunges into the Anthony Creek Trail. Now it is all downgrade to Cades Cove and he reaches that goal, another four miles, in what he considers record time. It is 8:00 P.M. when he reaches the ranger station in the cove and reports Rick Callahan missing. It is too late to even mount a hasty-search, so Walt Johnson is interviewed in depth for information about Rick and plans are made for a full-scale search tomorrow. These include a hasty-search sweep of all trails in the vicinity of Point Last Seen (PLS) and off-trail search teams following the stream courses in the Bone Valley drainage.

THE JOURNAL

Friday, May 10, 1985 6:00 A.M. Day 4

Having been called early in the previous evening, I arrive at the staging area in Cades Cove before dawn and get some coffee while assignments are being determined. Finally, about 7:30, several searchers, 3 dog teams, and I leave Cades Cove by jeep for Spence Field. About a mile up the jeep trail one of the jeeps flips into the creek throwing dogs and searchers about. No one is hurt; we right the vehicle and proceed to Spence. Another mantracker and I, with two dog teams, are taken by Walt Johnson to the place where he and Rich dropped off of the AT. As we walk down the trail, one of the tracking dogs, a very large black German Shepherd, takes off after a wild hog down Eagle Creek. Sensing a less than acceptable level of training for this animal, I tell the dog handler that if she can catch her dog she should remove him from the search area. We break into teams and dog handler McClure with his Border Collie and I proceed together down the slope to the left fork of Roaring Creek. I follow two sets of men's tracks into this drainage. The Collie is an excellent tracking dog finding tracks I would miss.

As we move away from the ridge crest and into the ravine the slope steepens to a precipitous drop and the underbrush closes in. Soon we are well into the drainage of a small stream, the headwaters of the west fork of Roaring Creek. By this time it is around noon and at first there is no sign, but as we reach an elevation of about 4120 along the creek, I find two bootprints going downstream which look like they are about two to three days old. At about 2:00 P.M., at elevation 3960, I find a single print still headed downstream.

At elevation 3600, I find a dark blue stuff sack containing a light blue sleeping bag which is wet and hanging out of the sack. Next to it is another stuff sack, red, containing only tent poles, no tent. Both of these items are lying on a log in the middle of the stream. Within 20 feet of this location, I find a red string which is a perfect match to the red stuff sack. Ten feet farther on, there is a clear footprint in the mud along the creek. It is from a Vibram sole boot about 12 inches long and headed downstream. It looks to be about two days old. The time is now about 3:00 P.M. and we hasten on. At elevation 3440 I find a scrap of orange poncho stuck in a bush next to the stream. This seems clearly to be a marker placed there by a person lost and either wanting to provide a sign for searchers or to guide himself back if he wants to retrace his own steps. Another searcher coming into our same search path from another stream course, reports on the radio finding more scraps of orange poncho downstream below our location.

Farther down stream at elevation 3400, at the junction with the East Fork of Roaring Creek, our tracking dog shows some interest in following that fork upstream. Based on the evidence of the direction of the tracks and the scraps of poncho, however, we believe that the search trail is still headed down and we continue in that direction. At elevation 3120 we find the rest of the orange poncho, a large piece, placed in the fork of a small tree in the middle of the old logging railroad bed. This time the dog shows interest in going in a westerly direction, but we decide to continue south on the old rail-road bed looking for more orange poncho or other signs of Callaban. It is now a little after 4:00 P.M..

Some distance farther along, at elevation 2720, we find in a muddy patch in the middle of the road bed, two footprints made by a tennis shoe headed downstream. This is a puzzle since we have been following a lug-sole boot. We are aware that Callaban is believed to have tennis shoes in his pack so we think that it is conceivable that this might still be Rick. The prints look to be about three days old. We continue to follow the road bed downstream , cutting for sign. By this time it is 6:30 P.M. and we are beginning to tire. From this point on for the next mile or so to Hall's Cabin we find no sign of Callaban. We do find tracks of another biker, probably Johnson whom we know hiked down to the cabin looking for Callaban.

We spend the night at Hall's Cabin at elevation 2480 and develop a plan to continue the search the next day. We have been reporting our progress and information developed to search head-quarters by radio all day so there is no need to file a further report. We do relay and confirm our plans to return to the last location where we could be fairly sure we were still on the Callaban trail, where the remnant of the poncho was stuck in the tree crotch, and continue from there.

Saturday, May 11, 1985 6:00 A.M. Day 5

The combined search teams, which all arrived here by following different tributary streams in this drainage, all move together back up to the poncho-in-the-tree location. Altogether we are seven searchers, one dog, his handler, and me, as the tracker. As we proceed back upstream and reach the junction with Desolation Creek at elevation 2800, our dog again alerts and wants to follow that course. We flag the location for later investigation but continue on Roaring Creek headed upstream and spread out on both sides of the stream cutting for sign.

As we reach the poncho tree location, one of our group spots poncho scraps headed west. We follow this trail about 200 feet west and come upon a campsite where someone, probably Callaban, has spent a night. The camp is at the stream and is hidden from the old railbed we have been following by a high bank and dense underbrush. At the camp we find a fire ring of stones from the creek, a stuff-bag containing a wet sweater, a Frisbee, a right tennis shoe and several food items. On a tree in the camp area is a note to Walt from Rick covered by a piece of the orange poncho. It says that he is lost and is headed north "to Knoxville." Digesting this information, we decide to again split into separate teams and follow different courses. Two rangers are to follow Rock Camp Branch to the northwest. The rest of us, with the dog, will continue north on the east fork of Roaring Creek where the dog alerted yesterday. By now it is about 1:00 P.M. and we stop for a quick lunch.

We hike past Mineral Gap Branch which the other team searched yesterday and follow Roaring Creek to the confluence of the east and west forks, then turn up the east fork. At about 3:00 P.M., we reach elevation 3520 and find two clear bootprints headed north. They are the same 12 inch Vibram lug soles we followed downstream yesterday on the west fork. They appear to be about two days old. Overhead we can hear the helicopters cutting back and forth looking for Callaban.

As we reach elevation 3800 feet we find an apple core, some uneaten fruit roll-ups and several carrots but nothing else. We continue climbing upstream. It is now 4:00 P.M. At elevation 3880 we find a pocket from Rick's orange poncho full of torn pieces which he had been using as markers in the middle of the stream. It is now 4:30 P.M. At elevation 3960 at 5:00 P.M., we reach a three-way fork in the stream course and split into three groups to follow all of the branches. At 5:30 P.M. at elevation 4080, a ranger on the eastern branch reports that he has found a large container of drink mix, some tomato paste, and some other Italian food items in a zip-lock bag. We all join him and continue up that drainage.

It is now 6:00 P.M. At elevation 4120, we find a scrap of a back country trail map of the Smokies in the middle of the stream. At elevation 4200, the trail of occasional fragmentary boot tracks we are following leaves the soft ground along the stream and wanders about 20 feet off to the right. We can find no sign of the track beyond this point as we continue to climb the ravine, cutting the banks in the gathering dusk. At about 7:30 P.M., we top out our climb on the AT just east of Thunderhead Mountain. From here we head out to Spence and transportation by jeep to Cades Cove.

There is much talk about what might have happened to the track we were following and a lot of speculation about Rick and his current condition. We know that he is without his sleeping bag and sweater, has lost or thrown away a lot of food he was carrying, is probably wet from the creek and the periodic rain storms we've been having. The signs we have seen as we track him raise ques-tions about fatigue and hypothermia either of which could be fatal if he became sick or injured.

Sunday, May 12, 1985 8:00 A.M. Day 6

The sun is shining as we leave Cades Cove by jeep for Thunderhead peak. The plan for today is to go back to where we lost the track yesterday and search in all directions to try to pick it up again. Helicopter searches will be focused east of the east fork of Roaring Creek and Mineral Gap branches. Ground search will also be directed toward this area.

Moving back down what is now a pretty familiar area, we are heading toward the location where the last prints were seen when we hear over the radio that a ranger in a helicopter has spotted someone on a ridge west of Calhoun Branch and south of Brier Knob, about a mile east of our location. From the description, it is likely to be Callahan. The ridge is too steep and the underbrush and

trees too heavy to land the helicopter, but it hovers to guide
searchers on foot to the location where the lost camper has built a
signal fire and is waving to spotters. It takes some time for searchers
on foot to reach the man and when they finally do, he runs away
and has to be gently restrained. It is Rick Callaban and at first he
is wary and will not talk to his rescuers. However, there is one
woman in the group of people who are assisting him and after a
while he begins to talk tentatively with her.

They feed Rick and give him clean, dry clothing. He seems to be
a bit haggard and has minor scratches and bruises, but appears
otherwise to be healthy. At about 2:30 P.M., as they start to walk
back toward the AT, which is in fact less than a half mile away
from the location where he was found, he becomes more and more
garrulous and euphoric. At about 5:00 P.M. the rescue team with
Rick Callaban in tow reaches Brier Knob and the AT. From here
they hike the short distance to Derrick Knob where all are lifted out
by helicopter at about 7:30 P.M.

While all of this is happening, we climbed back out to the AT
and hiked east to join the rescue group at Rick's location. Once
among the group of searchers who first reached Rick, I had a
chance to talk with him. At first he was very wary, but as he began
to warm up he confided in me about the scary sounds he heard
every night while he was camped alone. One was a whoosh sound.
I asked Rick where he had urinated. He pointed to the area. I asked,
"Is that where the whoosh came from?" He replied, "Yes." I told him
it was deer after the salt he had left them. Deer mice were the
rustling sounds he heard, and large owls were the sounds coming
from the trees. Wild hogs and bear accounted for the other sounds.

On May 14, another ranger and I were assigned to backtrack
Rick from the point where he was found to try to determine how he
got from the place we lost his tracks to the place where he was
found. At the "found" location we cast criss-cross patterns around
his makeshift campsite until, at elevation 3600 feet, we picked up
his footprints coming from the west. We followed these down to elevation 3400. Just below this point we lost the tracks. However, from
my interview with Rick two days before and from the physical evi-

dence and from my general knowledge of the area, I think my reconstruction of the missing three days of his experience is reasonable.

5/10/85

Rich awakens having had very little sleep. Hearing helicopters in the area to the west, he waves frantically to no avail because they are too far away. He spends another day sidehilling to his right, crosses a small stream and climbs another very high ridge where at 2:00 P.M. he again gathers vast amounts of firewood to ward off the night's monsters. Rich has no one to talk to. He becomes even more apprehensive finding himself, at times, running in a blind panic at a forest sound he does not understand. This night is even worse than the previous one with new sounds of grunts and groans and howls. He asks himself, "Am I losing touch with reality?" He does not know it but he is suffering from a combination of hypothermia, fatigue, and panic/depression, a result of his fear.

5/11/85

Rich awakens to what is now a routine of eating, walking, resting, gathering wood, and trying to sleep. This day he continues sidehilling, crossing a creek and still heading to his right. He climbs a very rugged ridge in hopes of finding a clearing and waving to one of the occasional helicopters that he now believes are searching for him. But Rich only encounters more dense brush. After walking most of the day Rich again stops on a high ridge that is more open than the others he has encountered. He sets about gathering wood for a fire. After building a fire, he writes in his daily journal a letter to his family expressing hope that his log will someday be found so they will know what has become of him. He gets out his first aid kit to doctor his sore body and finally notices a book which he has had all along giving instruction as to what to do if lost in the wilderness. It says to stay put, build a shelter, build a signal fire, etc. He is relieved enough to think that he may be found if he follows these instructions. This night is no less frightening than previous ones but be manages to cope.

5/12/85

Rick awakens with the idea of building and maintaining a smoke signal, which he does. Helicopters are again circling around to the west as the fog lifts. He throws green material on the fire which is now roaring hot and great clouds of smoke billow out. Finally, later in the morning, one helicopter comes to his smoke and circles overhead. He waves frantically and they drop him a packet containing food and instructions to stay put, searchers are coming to him. He does not know what else to do. He is very anxious and confused. At 2:00 P.M. Rick is frightened to hear the breaking of brush and something coming toward him from up the ridge. Men come breaking through the brush at the edge of his camp. Rick runs from his camp in full flight away from the sounds. He looks back to see a man has caught up to him and is restraining him. He is taken gently back up to the camp where there are other searchers. One is a woman and he talks to her as they walk out. Walking up the ridge he gets to feeling better about being around people again. He feels good now and he talks their heads off about everything. After they reach what looks like a real trail people give him food and dry clothing to put on. Rick hikes out the trail with the people and at Derrick Shelter is flown out via helicopter. It takes some time for Rick to reorient but it helps when his parents greet him later.

The social psychology of human beings in the wilderness can often be deduced from their behavior as read from the evidence they leave behind. This is the most interesting part of the art of "tracking." From the relationship of two sets of tracks, for example, it is easy to tell who is the leader, or "alpha" individual and who is the follower or, "beta" person. Commonly one set of tracks will always be first, sometimes close to, but almost never behind the other. Alpha will go ahead, will explore away from the chosen path or direction and will show other signs of independence. Beta will always follow, will stand with feet closer together and will tend to stay put or even return to a previous location rather than venture away. From the tracks of Walt and Rick and from what we know of their story in this search, I believe Rick to have been a "beta" who would try to return to areas where he had been before.

This is why he tried to backtrack through what he knew was very difficult terrain rather than go ahead. Hamlet must have been a "beta" too!

People who are lost off the trail system will also tend to move in certain systematic ways as well, even when they might appear to be wandering. For example, when faced by a barrier such as a stream or cliff, they will almost always turn in the same direction. For some it is to the right, for others to the left. This is useful to know when casting about to find the continuation of a broken trail. If I have once determined the pre-disposition of the person I am tracking, I know where to look first when I lose his tracks. I felt that Rick had a right dominance because while trying to self rescue he always went to the right.

Rick was lost or alone from 5/07/85 to 5/12/85 a total of 5 nights. I find, in analyzing the searches I have been on, that if a person is lost for over two days reality becomes distorted. Nearly always, the person located will hide or flee from his rescuers as Rick did. The psychology of being lost has taken hold of them. They only know fight or flight reactions. On the other hand, if a lost person has a companion, whether animal or human, they have a tendency to go toward their rescuers.

This search is typical in that I hardly ever actually am the first to come upon the lost person, but my skills are helpful in determining direction of travel to narrow the search area and to focus the efforts of other searchers. All searches are a team effort.

TWO DOG NIGHT • SHANE JUSTICE

The call comes in to the sheriff's office in Haywood County, North Carolina about 9:00 P.M. on Saturday, September 16, 1989. Denise Justice on Hemphill Road in Maggie Valley, just at the edge of the park, reports her son Shane, aged six years, missing from their home. Shane was last seen by his mother riding his bike in their driveway about two in the afternoon. The family searched for him for a few hours before reporting him missing when darkness was approaching. Mrs. Justice tells the sheriff that their two dogs, Shane's best friends, are also gone and are probably with him. Shane is wearing shorts and a tee shirt. She says Shane has always wanted to visit a rock outcrop that is across the valley from his home and she believes he may be there. The weather is variable with rain storms heavy at times and a high in the low seventies and a low of 40 to 50 in the mountains. It is clear and there will be a full moon.

Hundreds of searchers, including neighbors, sheriff's deputies, rescue squad personnel, state rescue teams, forest service and park service personnel search all Saturday night for the barefoot boy with his dogs. There is real concern for the boy who has no protection from hypothermia because he is only wearing short pants and a tee shirt and has no coat or other protection from the low night temperatures and the rain. Shane is not located but a group of rescue squad men do find a good footprint in the mud 4.5 miles north of the boy's home near the edge of the park. They rope off the area and are camped out at the track, waiting for assistance all night long. Searchers on all-terrain vehicles and on foot are combing the mountain. The elevation of Shane's home is about 4200 feet and the track was located at 4500 feet. This area is a steep mountain slope with meadows and heavy forest.

The search progressed for six hours Saturday night and the little boy had not yet been found. My supervisor, Duane, listening to the traffic on the radio, calls the search manager in the sheriff's office and tells 'im, he says, "We've got a guy who'll find him for

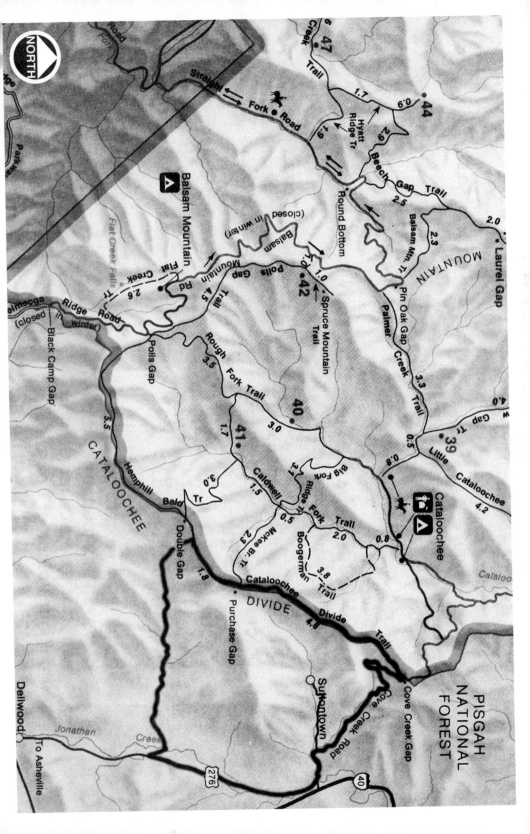

you. If you get to where you're gonna' give up, we got a guy that'll find him for you, if you're interested." And they said, "We're interested." That's when I get the call.

THE JOURNAL

Sunday, September 17, 1989   2:00 A.M. Day 2

It is still raining off and on when I receive a phone call at 2:00 A.M. from Joe Smith, south district ranger in the Smokies. He relates the story and says they are camped on a track which they want me to follow. I tell him I will be there as soon as I can, probably about 6:00 A.M. I arrive at Maggie Valley at 5:30 A.M. and encounter a deputy blocking the road at the bottom. I'm driving my vintage 1970 Ford station wagon which I purchased for 200 dollars not long before. That car had problems with the gas feed sticking and earlier in the week I was going home from work and a motor home passed me because I was going too slow. After the motor home passed, the throttle on the Ford stuck and I found myself going from 30 mph to 70 mph. As I was frantically trying to un-stick the pedal I came upon the motor home and had no choice but to pass. As I passed, I noticed a frightened elderly couple looking at me from the motor home. After completing my pass, I noted a sharp curve ahead so I pointed the wagon for a bunch of saplings on the shoulder whereupon I cleared a path through them. I managed to get into neutral with my motor racing at full throttle. The elderly couple drove by and gave me a strange look. Back on the road after hasty repairs I again came upon the motor home. I could see the same frightened faces staring out the rear view mirror. They pulled over and let me pass. I was going 30 mph.

The deputy on guard at the foot of Hemphill Road stares in disbelief at me and my vehicle when I tell him I am the "hotshot" mantracker who is here to find the missing boy and he only lets me pass after calling the search leader on the mountain. I proceed up this very steep grade and arrive where there are lots of rescue vehicles parked on the road. As I park my vehicle and get out, the driver's side door falls off its hinges onto the middle of the road.

Bam! It hits the road, and I get out and try to tie the the door back together. I have a rope and I tie the back post of the front door to the front post of the back door, and then the door'd swing out from there; so I tie it to the rearview mirror then, and it stays. Finally I get my bag out.

After tying my door back together I procure a couple of sausage biscuits and hot coffee and go to the hood of a truck to talk with the leader. He looks a little doubtfully at me but then he says a jeep will take me to the track soon but could I suggest assignments for the hundreds of searchers waiting? I tell him it would be good to look under, and around the many summer cottages in the area and walk all roads looking off the shoulder. Also some folks should wade all streams looking under everything there.

It is a bumpy ride up the mountain but soon enough we arrive at the park boundary. Along this part of the boundary, the park service has constructed a trail which follows the boundary, demarking that line. There are clear markings on the trees and at every change in direction a monument. There are also signs posted to notify those entering the area that it is park property.

As we walk northeast on the boundary trail past Purchase Gap, searchers are on horseback and in or on any vehicle you could imagine. It took several hours to get to the men who were camped patiently on the track. I have the highest praise for these men. They had erected a tarp over the track and had diligently protected it from rain and human and horse impact.

The footprint was in a little muddy area. It looked like it would be the boy, because there are two dogs with him. One dog is out roaming around in the woods, another dog is right beside him on his right side all the time, like he was holding him or something. His mother said that he left with his dogs so we are almost certain now that this is Shane. Once again I wind up tracking the dogs. We fol-lowed the one dog that was always next to the little boy right up to the park boundary. Where the boundary trail went right he stopped at a little gap about half a mile from where we first saw his track and turned straight down into the park. Because the other dog that was roamin' free went that way, the boy just followed.

I follow the tracks into the headwaters of Den Branch into the Great Smokies National Park whereupon I tell them by radio that the boy has entered the park and is proceeding deeper into it. His bare foot print is approximately five inches long and corresponds well to a six year old's track. If Shane continues in the direction he is headed, it will take him down the drainage of Den Branch, eventually crossing the Boogerman Trail which follows Caldwell Fork to which Den Branch is a tributary.

Tracking is slow work and whoever you're following is usually moving faster than you can. A lot of times I don't actually come upon the lost person because I can't move as fast tracking. Searches are a team effort and without the people that run the trail looking for signs or those assigned to intercept the trail I am tracking I don't believe I'd be able to operate. I always tell the other rangers that do run the trails to walk to the side of the trail, never walk in the trail, because if you walk in the trail you're gonna' destroy something, I don't care how much you care how much it rains, if you don't damage the sign, there'll be somethin' left there. And the steeper slope you get on, you're gonna' slide and your footprint will still be there after a gully washer. However, even if you can't easily catch up, if you can get a bead on the direction and more searchers to intercept that course, you can have success.

Following that strategy, I request that a particular ranger from Cataloochee whom I know would know the Boogerman Trail be assigned to go up that trail to Den Branch from the bottom and proceed upstream on the branch to try to intercept the boy. As I was tracking him, I noticed that one dog always stayed to the boy's right side, never leaving him, but walking beside him. I felt the boy was holding onto the dog in some fashion. The other dog would go hither and yon and made my job harder as I would not proceed until I tracked each party out.

At elevation 3400 the boy turned uphill again and to the right, heading toward Panther Spring Gap. This information is also radioed to search management and, based on this information, more searchers are deployed along the boundary trail northeast of Panther Spring Gap to search the woods north of the trail. We continue on Shane's trail climbing the slope and there at the top of the ridge is a place where he spent the night under a big log; and the dogs were in there with him. The area is all scuffed up and there are dog tracks everywhere.

And he is starting back, because the dogs are leading him back towards the boundary trail. At this point these searchers on the ridge above me hear Shane singing in the woods near the boundary trail. They call to him and the boy goes to them. He is cold, but not really frozen because he has his two dogs. He slept with those dogs all night. He says, "My dogs got on either side of me and kept me warm." But the fleas have eaten him up! Poor little feller. He has flea bites behind his neck and ever'where. When the fellers approached him, the boy did not run. The dogs ran but stayed nearby, circling us. But the boy didn't run because he had two companions, his two friends. Then I came on up the hill and met them. We put dry clothes and socks on Shane and fed him. I ask him if he had eaten and he replies, "Only berries". He says that one of his dogs had caught a chipmunk but the dog would not share it. We carry the boy on our shoulders out to the boundary trail and then northeast toward Cove Creek Gap where we are to meet a vehicle which will transport Shane out. On the way we meet a group of horseback riders who volunteer to help. I ask if one of them would get off and let the boy ride while they lead the horse. Most seem reluctant to do so, but one lady agrees. As we proceed to Cove Creek Gap the horse riders start to turn off into the woods along with our lost boy. I stop them and ask where they are taking my boy. They say, "This is a shortcut." I tell them, "No way, this boy has had enough shortcuts for today." So the lady comes with us leading the horse with the boy on top. The others take the shortcut.

As we arrive at the Cove Creek Main Road, I am in the rear. The sheriff and many searchers are there along with a herd of media. One of them sticks a microphone and camera in the lady's face that was leading the horse and says, "Tell us how you found him." "Well, I let the lady fill them in as my job is done. I do think about how the little boy dodged hypothermia and stayed warm during the night by getting between his dogs for warmth. The only drawback to this was *the numerous flea bites around his neck. He actually could have done well with only one dog as the temperature was only 40 during the night. This would be a positive statement for taking man's best friend with you in the wilderness.*

Old timers will tell you that a one dog night is 32 degrees and above. This is to say, one dog on top of you during the night. A two dog night is around 20 to 32 degrees. This means, a dog on either side of you. A three dog night is below 20 degrees, when you need one dog on either side and one on top!

I wonder if the singing group Three Dog Night got their name this way?

And you know, right after this last boy that I found, I went into a restaurant near there after the search and we ate dinner and I went to pay the checkout lady and she says, "The manager says your dinner is free." I said, "Free?" She says, "Yeah, it's free. He says your money's no good here!" I said, "How come?" She says, "He's seen you on TV where you found that little boy." I says, "Well, how many times do I get to eat free? Is it for a week, or what?" "No, just this one time," she says.

AFTERWORD

LESSONS FROM THE STORIES

These are not freak accidents, but typical stories about the experiences of wilderness visitors which occur every day somewhere in our national parks and forests.

Avoid the edge!

If you are a hiker, a few simple rules drawn from these stories will reduce the chances of you being another story in the media or another statistic in the official records. As you read these words, they will raise echoes of the stories you have read. Think of what might have been.

DON"T GET LOST

Avoid Disorientation

Always carry a map. And use it! A compass helps too!

Trying to find yourself on the map <u>after</u> you are lost is not very effective. The best course is to use your map continuously as you hike, keeping track of local landmarks and topographic features as you go. That way, if you go astray, it will soon be apparent to you and you can retrace your steps to the last familiar spot. Share the "orienteering" activities among all members of the group so that everyone maintains their orientation and can act intelligently, even if they somehow become separated from the leaders. Every member of the group should have a map. No matter what else you do or do not do, avoid leaving the trail system. If you are believed to be lost, the first thing which will happen is a "hasty-search" of the trails. If you are someplace on them, even away from your intended route, you will be found, and quickly. It is for this reason that cross-country hikes should be avoided except under the most exacting conditions.

Safety Margin Gear

The equipment you carry should be more than adequate for the intended activity. If you can reasonably expect cool weather, prepare for cold! Even if the forecast is for sunny skies, carry rain gear, hat, etc. When it comes to sleeping bags, pack frames, rain gear, etc., always buy the most effective equipment you can and keep it maintained. The edge is always closer than you think and an extra margin for error can mean the difference between life and death. Most important is a good first aid kit with an adequate instruction book which you should read before you need to use it.

Buddy System

It is not wise to go off into the wilderness, even on the trails and even if only on a day hike without a companion. Even the most experienced and fit hikers have accidents and can be injured. Children should never be allowed to wander out of sight of adults and should be taught these same rules, including "orienteering", before they are taken into the wilderness for the first time. If they are too young, do not take them.

Itinerary

Not only do most federal and state parks and forests require it, but it is simply good practice to file a trip itinerary with the relevant agency and with friends or relatives. The knowledge that you will be missed and a search will be initiated quickly allows anyone in trouble the time necessary to protect themselves in place, rather than plunging over the edge in a futile attempt at self-rescue.

Avoid Exposure / Hypothermia
Be Season Wise

Carry and use adequate clothing and protective gear so you can maintain your body temperature and keep your extremities comfortably warm. Learn what kinds of gear are appropriate and necessary to do this for the season you will be hiking. Even the early physical stages of hypothermia are to be avoided because they very quickly and subtly lead to the psychological effects which rob you of your strong self-preservation motivation and can lead to hypothermic anomie (disoriented thinking), and eventually, death.

Know The Weather Forecast and Act Accordingly

If extreme weather conditions are expected, postpone your trip. If you encounter unexpected conditions, don't forge ahead, but remain in a shelter until conditions improve and you can safely proceed. Do not delay in donning foul weather gear or building a fire if you are caught under worsening conditions. Whatever you do, do not leave the trail system even if you are sure you know a "short cut" to shelter or your car.

IF YOU ARE LOST ANYWAY

Do not panic. It is far more important to take the time to stop, get your bearings if you can, assess your situation, make a rational plan for action, and proceed deliberately. It is rare that quick action is needed and, in fact, hasty judgments and frantic physical activity more often lead to worsening the situation rather than resolving it. Here's what to do:

If you are on the trail system, but disoriented:

Do not leave the trail. If you recall a trail intersection, shelter, or other landmark feature which was recently passed or seen, return to it and get your bearings using your map and compass.

If you have lost your map/compass or if it is night or bad weather, try to find some shelter adjacent to the trail, gather wood, build a fire and settle in to wait for improved conditions. Do not waste your energy and stamina on needless activity. Conserve your energy and body heat by protecting yourself as much as possible.

If you have somehow lost the trail
and are not sure how to regain it:

If you have your map and compass and some idea of your approximate location and orientation, try to decide if a trail is close above you. If so, climb up to it being sure to follow a compass bearing so you do not wander in circles or random segments. Since climbing is more physically taxing and risky, you should only attempt this if you are relatively certain that it is the correct thing to do. A good example is in the Great Smoky Mountains, where the Appalachian Trail follows the ridge line and going uphill will bring you to the main trail.

If you are not sure that a trail is close above you, then your best course of action is to either go downhill or stay where you are. If you are in dense woods and/or if it is unlikely that you will be missed and a search initiated, then you must hike down because you are unlikely to be searched for and found under these conditions. If you follow a stream course down stream, it will usually eventually lead you either to a trail and signs directing you, or directly to civilization. Hiking down is much easier than up and you will make greater progress toward self-rescue before you become exhausted or incapacitated. Obviously this course of action depends upon the local conditions. If you are in a wilderness area where your map indicates that downstream leads away from civilization, do not proceed in that direction.

If you are injured and without assistance:

First assess how serious your injuries are. If you are losing significant amounts of blood, stop the flow immediately. If you are physically incapacitated to the point of immobility, try to make yourself as comfortable as possible and protect your body from exposure. Build a fire if you can and wait for rescue if you can reasonably expect a search to be launched. If you have followed the other rules, you can expect help soon.

If you are in a relatively clear area and know that you will be missed, stay put. Make a camp of whatever kind you can, be sure not to become dehydrated, take care not to lose your body heat and your energy reserves. Build a fire, get comfortable, and wait for help.

ABOUT THE AUTHORS

ABOUT DWIGHT MCCARTER • BACK COUNTRY RANGER

"I believe I've made a difference," says Dwight McCarter. And he has. His 27 year career which began in 1967 with the National Park Service in the Great Smokies is punctuated with heroic search and rescue missions as well as other accomplishments.

Dwight McCarter was born in 1945 in Townsend, Tennessee and raised in the Smoky Mountains area of Blount County. His family home is located less than a quarter mile from the Great Smoky Mountains National Park boundary and only a few feet from the Little River. The family history goes back to the first settler in the Gatlinburg area. He absorbed the lore and love of the mountains through hunting and fishing trips with his father. Few know and love the Smokies more than Dwight; they are his legacy and his "best friend."

After completing high school Dwight spent some time in army service before beginning his park service job which would range from picking up trash on hiking trails to apprehending criminals in the Great Smoky Mountains National Park. When he became a back country park ranger his tasks ranged from fire fighting to man tracking. He also explored and documented vast and remote areas of the Smoky Mountains. Numerous commendations for search and rescue missions and countless expressions of appreciation for the sharing of his vast knowledge of the region attest to the high regard in which he is held.

Dwight's skill, strength, and endurance as a mountain hiker and back packer are legendary. In September 1978, curious about the tall tales of some of the hikers he met, he decided to take on the challenge of hiking the full length of the Appalachian trail from Fontana Dam on the southwest to Davenport Gap on the northeast. The distance is 72 miles over terrain which varies from about 1500 feet elevation at Fontana to over 6500 feet at Clingmans Dome before dropping again to below 2000 feet at Davenport. Between these points, the trail follows the ridge line of the mountains, continuously climbing or dropping the entire distance.

Dwight left Fontana at 5:00 A.M., woke a sleepy camper at Mollies Ridge Shelter at six to sign his verification log, crossed Thunderhead Mountain at 7:15 , a distance of 16 miles. At 9:30 A.M., he met a trail crew at Silers Bald with another 13 miles behind him. He was averaging 6.4 miles per hour through some of the most demanding terrain in the eastern United States. At 11:30 A.M., he reached Newfound Gap and stopped for lunch, hitting the trail again within the hour. By 2:30 in the afternoon he had reached Pecks Corner where he encountered a skunk who refused to yield the trail. Despite a detour to avoid the recalcitrant critter, Dwight reached Tricorner Knob by 4:00 P.M. when it started to rain a cold, drenching downpour. Knowing the serious effects of hypothermia, he decided to terminate his effort and dropped off the trail at 7:30 P.M. at Cosby, just a few miles short of the full distance of 72 miles. In that same year, Dwight logged almost 1300 miles hiking the back country trails and "hells" of the Smoky Mountains.

Now retired from park service, Dwight still lives in his hometown with wife Sally and son Rob. He continues to be fascinated with the Smokies, with Indian petroglyphs and pictoglyphs, and with early settler lore. He conducts popular tours and presentations throughout the area.

ABOUT RONALD G. SCHMIDT

Ron Schmidt was born in 1931 and raised in New Jersey. He attended Columbia College, then completed his M.A. and Ph.D. degrees in Geology at Columbia University and University of Cincinnati. Now retired from teaching and administration at Wright State University in Dayton, Ohio, he continues a long standing professional and business career in geology, ground water management, and administration. Writing and publishing are the latest additions to his varied career. He lives in Yellow Springs, Ohio with his wife, Phyllis.

Ron became interested in the Great Smoky Mountains while conducting engineering geology field courses there in the late 1970's. The clear indications of past industrial activity within the park boundaries piqued his curiousity and fed a long-smoldering interest in the history of the area, particularly the brief, but critical logging era. He began collecting information on the logging and railroad history of the mountains, and tracing the location of the railway roadbed there. His book on the subject, *Whistle Over the Mountain*, was published in 1994 by Graphicom Press. It has received critical acclaim from book reviewers and regional historians.

While working on the manuscript of Whistle Over the Mountain, Ron met Dwight McCarter, whose detailed knowledge of the mountains enabled Ron to locate important historical features. Dwight's stories of his experiences as a park ranger and his knowledge of mountain lore seemed destined for another book. Together, Ron and Dwight decided to publish Dwight's personal journal of his experiences in search and rescue as *LOST!*.

Another Graphicom Press regional title which may interest the reader is **Whistle Over the Mountain**. *Whistle* reveals the little-known history of the Great Smoky Mountains National Park when timber was king and the forest resources of these magnificent mountains were widely harvested for human use. This book with over 150 historic photographs, 40 original pen and ink drawings, detailed location maps and chronologies is available now. Ask for it at your favorite bookseller or order directly from:

Graphicom Press
P.O. Box 6,
Yellow Springs, Ohio 45387.
(937) 767-1916